The Tar Sands

The Tar Sands

Syncrude and the Politics of Oil

Larry Pratt

Hurtig Publishers
Edmonton

Hurtig Publishers
10560 105 Street
Edmonton, Alberta

ISBN 0-88830-098-0 cloth
ISBN 0-88830-083-2 paper

Printed and bound in Canada
By John Deyell Limited

Contents

For Tricia

"There is a whirring of wheels, a hissing of steam, and three things emerge from that same machine. One is water—the natural moisture content of the sand plus the water that has been used to emulsify the mixture; the second is sand—pure, white, sparkling sand; the third is a black stream of valuable hydrocarbons containing naptha, fuel oil, lubricating oil, kerosene, asphalt—all ready to be passed on to a refinery for cracking into the various valuable components."

The Edmonton *Journal*
November 2, 1929

"We want you in on what we're doing."

Syncrude Canada Limited

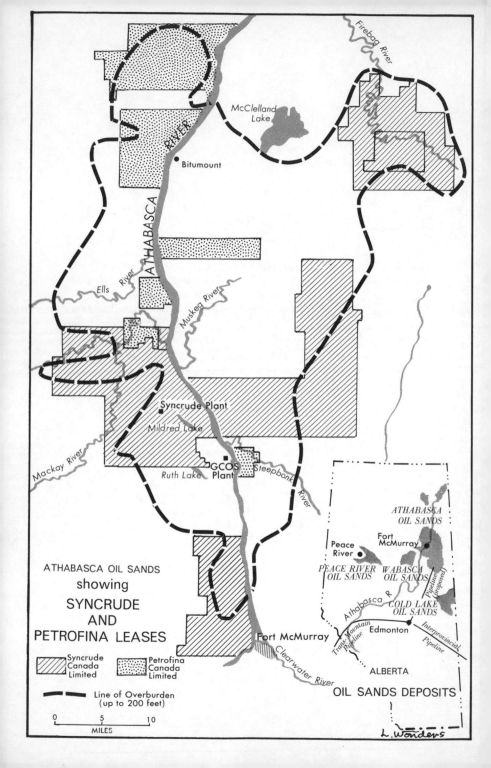

ATHABASCA OIL SANDS
showing
SYNCRUDE
AND
PETROFINA LEASES

Syncrude
Canada
Limited

Petrofina
Canada
Limited

Line of Overburden
(up to 200 feet)

0   5   10
MILES

OIL SANDS DEPOSITS

Preface

Anyone so foolhardy as to write a book about energy in Canada in the 1970s is risking the likelihood that his or her research will be overtaken by events and rendered obsolete before it reaches the bookstand. This was painfully revealed to me in December 1974 when Atlantic Richfield Company, one of the corporate members of Syncrude Canada Limited's oil sands joint venture, withdrew from the project, leaving both the Athabasca sands and my finished manuscript on the resource high and dry. ARCO's rude behaviour initiated a lengthy episode in the politics of Canadian oil that concluded two months later in a hotel room in Winnipeg with the decision of three governments to rescue the Syncrude project with huge infusions of public cash. Although these events only confirmed my original conclusions about what was happening in the tar sands, I decided that they were too revealing to go unrecorded and unexamined. And even though this book has gone through numerous changes since the first draft was written, events are still unfolding too swiftly in the tar sands and elsewhere for a definitive work on the subject to be produced.

Nevertheless, I hope that this book, which is as much about the politics of oil as it is about the tar sands, will retain some value in helping the reader to analyze future developments in Canadian energy. The essential thesis is that Syncrude is likely to become the prototype for new energy ventures in Canada; and if this is correct, every Canadian should know something of Syncrude and of the remarkable power of the oil lobby in our political system. If power is defined as the ability to realize one's will and to achieve one's objectives, then the oil lobby necessarily must be reckoned as one of Canada's fundamental power blocs. The documents which are dis-

9

cussed in chapters eight and nine illustrate the oil industry's easy access to the highest levels of political leadership in the country, and confirm its strong influence over vital energy decisions—pricing, taxation, environmental regulation, labour legislation and other policies struck at both the federal and provincial levels of government. Indeed, anyone who attempts to reconstruct Syncrude's lobbying activities in Alberta and in Ottawa from 1972 to 1975 will be hard put to find a single issue of any substance where the owning interests in the consortium failed to win their basic objectives. This private power—and the absence of any countervailing power representing the public interest—is a phenomenon which most analysts of Canadian politics, who begin with the assumption that the state enjoys the last word in power, conveniently ignore. This book starts from a different assumption, one that unfortunately is taken generally to be very outdated; namely, that it is the job of the social scientist to try to explain how the real world works.

I should like to thank those who assisted me in gathering information on the tar sands—much of which is not yet easily accessible in the published literature. Because of the controversial nature of the opinions expressed herein, it is not possible for me to name those public and private officials who generously gave me their time and advice. The present political atmosphere in Alberta is such that criticism tends to be regarded as treasonous ("alien forces," to quote Premier Peter Lougheed) and unpleasant facts are dismissed as ideological heresy. One can hope that the day will soon come when those who hold so much power will feel secure enough about themselves and what they are doing that they will feel free to debate the issues in the open, spirited manner which is vital to the health of democratic politics. Until then, let me at least express my gratitude to the following: Tricia and Jeanette Smith, John Richards, Ed Shaffer, Jim Anderson, Jim Laxer, Randy Morse, Marianne Lindvall-Morse and Mel Hurtig.

The book is arranged as follows: chapters two and three introduce the subject of the tar sands and provide an historical overview of the resource. Chapters four and five view the pressures on the tar sands against the background of recent shifts in U.S. energy policy and Canada's tradition of developing resources along continental lines. Chapter six looks at the structure of the international oil industry and assesses its interests and objectives vis-à-vis the tar sands. Chapter seven attempts to caution against the policy of too rapid exploitation of the huge reserves of petroleum known to exist in the tar sands.

Chapters eight, nine, ten and eleven deal with the politics of oil in Canada, particularly as these are revealed in the bargaining over the terms for Syncrude. Finally, chapter twelve points up some of the long-term implications of the Syncrude episode and argues the need for public ownership and development of the tar sands. No doubt some people will find this alternative too radical for their liking. "Radical" literally means "going to the roots," and under this definition of the term I plead guilty to being radical, for it is the oil industry above all else that is at the roots of Canada's energy predicament.

1   Impressions of the Future

"It would be well to bear in mind that the present of today was the future of yesterday and that it is what it is because of the human actions, the human decisions of yesterday. Therefore the future will be what we make it."

John W. Dafoe

Fort McMurray seems an unlikely place to go looking for a glimpse into Canada's future. The long, 265-mile drive from Edmonton's modern skyline and urban sprawl northeast to McMurray, lying at the confluence of the Athabasca and Clearwater rivers in the heart of the Athabasca tar sands, feels more like a journey backward in time to a remote, faintly remembered pioneering past. Near Edmonton the countryside has been tamed, the bush rolled back and the land brought under cultivation; here man holds sway against the wilderness and the hard subarctic climate. About 100 miles out of Edmonton, however, as the car swings north onto Highway 63, the features of the land show faint evidence of man's imprint. Only the lonely strip of highway, some pockets of road construction and the lights of an occasional oncoming car, hellbent out of isolation for the city, distinguish the muskeg-topped land and the endless boreal forest from what it must have been long before the arrival of the first white explorer in western Canada. That this tranquil, untouched wilderness is marked as one of Canada's prime targets for future heavy industrialization seems completely unreal. Yet it is already happening. A significant part of Canada's future is now unfolding here in the isolated forests and river valleys of northeastern Alberta.

Nearing Fort McMurray, just before the road dips sharply down into the flood plain of the river valley, the land is not so much

13

tamed as it is devastated. Scrapers, bulldozers, heavy trucks and other weaponry of forced growth are everywhere, gouging out new "industrial parks" and mobile home camps to hold the overflow of unhoused workers and their families. Tent cities and trailer parks sprawl around the fringes of the town; McMurray is literally bursting at the seams. Accommodation is so scarce that the local newspaper runs ads promising a $500 reward for information leading to the purchase of a house. In mid 1974 finding space to park a trailer meant adding one's name to a waiting list of 300 people. The town's future is uncertain. Property speculators have driven land values within the town out of sight, and government has been forced to freeze land transactions and to take other extraordinary and authoritarian measures to keep the lid on a potentially explosive situation. Fort McMurray is booming, a great deal of money is changing hands, but the town is undergoing intense pressures as a result of the decisions of the world's great energy companies to bring the tar sands into production. Here the effects of rapid growth involving truly massive industrial projects, imposed on an unprepared, remote northern community, could create a nightmarish social situation. In McMurray and its environs live several thousand people of native extraction, and these descendants of the original inhabitants of the Athabasca country are being ignored in the frenetic pace of development. The native people of this region, like their brothers and sisters all over the country, are seeing only the underside of white Canada's cornucopia.

Part isolated northern community with its own unique traditions, part modern company town, Fort McMurray today is a study in fascinating contrasts, the new schools and quiet residential streets cheek by jowl with the tent cities and endless rows of trailer homes, the affluent oil men and businessmen rubbing shoulders with the boisterous drunks downtown at the booming taverns.

From McMurray the road north arcs across one of Canada's historic rivers, the Athabasca, near its junction with the Clearwater River. Driving along the west bank of the Athabasca, one recalls the earliest discoveries and explorations of the strange, oil-soaked sands, visible along the river bank, and the mind reels at the prospect of a dozen or more of the world's biggest strip mines and industrial factories bunched along the river in response to a growing gap in the energy supply and demand graphs of North America. "Where now the almost unbroken wilderness holds sway," predicted one of the first explorers of the tar sands, "industrial plants may arise and tall stacks dominate the landscape." The only way to get a real impres-

14

sion of what that could mean is to visit the tar sands now when development is just commencing. Seen from the comfortable distance of the corporate boardroom, government bureaucracy or university classroom, the pressures and hazards of development seem abstract and academic; but up close the stereotypes and pat solutions fall away as the immensity of the scheme begins to sink in. In the tar sands everything seems to happen on a giant scale.

Some twenty miles north of McMurray we arrive at the very heart of the Athabasca tar sands. The bituminous sands and similar heavy oil deposits cover some 20,000 square miles of northern Alberta, a vast area of virgin territory; but the prime leases marked for early development are located immediately north of the townsite on either side of the Athabasca in what geologists have designated the McMurray Formation. Here certain of the deposits are particularly rich in oil and near the surface, ripe for exploitation. At Tar Island, some twenty-five miles downstream from McMurray, perched on the elevated west bank of the river in the most improbable of natural settings, is the only oil sands extraction plant now in commercial production and one of the world's first large-scale synthetic fuel projects—Great Canadian Oil Sands, a subsidiary of Sun Oil of Philadelphia.

Just north of the GCOS site lies lease seventeen of Syncrude Canada Limited's $2 billion project, focal point of national attention in the winter of 1974-75. Originally a joint venture of four American-controlled petroleum companies, Syncrude underwent a metamorphosis behind closed doors in a Winnipeg hotel room in February 1975 and emerged as the hybrid offspring of three multinational corporations and three Canadian governments. Failing new political manoeuvring and work stoppages, the giant Syncrude project should go into production at the end of the 1970s. By that time, depending on future world energy trends and the willingness of Canada's politicians to underwrite, with public funds, much of the risk and cost of further development, other oil sands ventures now on the drawing boards may be under construction on the east bank of the Athabasca.

Visual impressions of "the world's first oil mine" at Tar Island are almost impossible to convey on paper. GCOS looks like it came boiling out of the imagination of some early surrealist painter with a fascination for monstrous earth-moving machines and smoke-belching factories—a celebration of the triumph of technology over wilderness. The very nature of the resource demands gigantism. Forcing the asphalt-like sands to surrender their thick, sticky oil and then transforming that oil to a marketable product is an expensive and appallingly dirty business involving large-scale technology, some borrowed from coal strip mining and oil refining and some invented especially for the tar sands. To be economically viable the process demands large economies of scale: everything, from removal of the muskeg and overburden to extraction and processing of the oil, must be done on a Brobdingnagian scale. GCOS belongs to Gulliver's land of giants, but it will be dwarfed by the plants of Syncrude's generation. The scale of the technologies employed, the mountains of materials handled every hour, the size of the mining pit, the sprawling lakes of polluted wastes—these are at once awesome and horrifying. After eight years of intensive development the GCOS lease is an ecological disaster—acres of black, scarred earth, hills of heaped soils, deep open pit mines, sulphurous stench—a kind of northern Appalachia, a biologically barren landscape. Some of the damage can be avoided, part of the land may one day be restored to something resembling its natural state, but much of the destruction is simply the price we imply when we argue that we must exploit the tar sands to meet North America's appetite for nonrenewable resources or to keep Canada self-sufficient in energy. Sacrifice of the environment may be inevitable if security of energy supplies is a national priority; clearly, there are policy trade-offs and dilemmas which admit to no easy, instant solutions. But the present strategy of energy development in Canada entails some very heavy costs and highly questionable benefits, and nowhere are these more apparent than in the tar sands.

Under the direction of the Conservative government of Peter Lougheed, Alberta is presently charting a course of economic diversification and industrialization as insurance against the day when conventional oil and gas reserves, the basis of the province's current prosperity, are depleted. In principle, one can scarcely fault the intention of fashioning a less vulnerable, more broadly based economy, and the idea has understandable political appeal on the prairies. Yet in practice, Alberta's evolving policy closely resembles that familiar species of development strategy so aptly termed "forced growth" by Canadian business journalist Philip Mathias. Forced growth is a process of artificially encouraging and hastening economic development through the use of generous public incentives and subsidies and a willingness by government to underwrite much of the risk and cost in return for little actual ownership and control. As Mathias has brilliantly documented, in an age of powerful multinational business provinces who play this risky game often find themselves impaled and bled white, the cost of one generation's hasty and grandiose policies mortgaging the future of its successors. This is the threat posed by events surrounding the tar sands; the difference is that the outcome will affect not just Alberta, but the country as a whole.

This is a book about the Athabasca tar sands, the interests that are attempting to bring them into production, and the benefits and costs of the development policies. In many ways it is a critical book which argues the urgent need for resistance to and reversal of these policies. The demands on the tar sands originate in the unstable, shifting politics and economics of world energy, in the investment priorities and bargaining strategies of the international petroleum industry, and in the policies of our own governments. Such demands have little in common with the needs of those most likely to be affected by development, and they have little to do with the best interests of Albertans and other Canadians in deriving maximum benefit from the use of their exhaustible resources.

This conclusion runs against the conventional, official wisdom. According to the barrage of reassuring and comforting press releases pouring out of the private and public bureaucracies, Canadians are getting a fair deal in the tar sands; government is in control and determining all features of development; and the problems likely to result from such development are well in hand. Public relations is a dismal substitute for the truth, however, even if the latter is less palatable than the interpretations of vested interests. It is the thesis of this study that Canadians are not receiving fair value for the exploitation of their resources; that it is the handful of multinational companies, holding almost exclusive leasing privileges in the tar sands, that are dictating the conditions for their development; and that the attendant social and environmental problems are threatening to get out of hand even in the very early stages. Many of the heaviest costs of what we are allowing to happen will be paid by our children and their children; and we will deserve their opprobrium if we do not soon demand some radical changes.

This book has been written for a broader audience than those who specialize in the study and practice of resource development. Part of Canada's present difficulty lies in the fact that the knowledge and details of crucial issues are monopolized by a few who evidently feel they have a divine right, indeed duty, to protect the rest of us from disquieting information. Neither the academic community nor the media are serving the public by acquiescing in this situation. Many academics clearly regard the attempt to exchange ideas and knowledge with the general public as undignified; the adoption of strong public stands on issues affecting the future of Canada is deplored as unprincipled demagoguery. The country's journalists, on the other hand, generally are not elitist, and my own limited experience convinces me that most are genuinely interested in critical debate of public issues. But good intentions are not enough: too often reporters and commentators rely heavily on press handouts from the corporations and governments for their information. That it is a difficult, tiresome, expensive nuisance to generate alternative information, particularly on highly technical issues, is undoubtedly true; but it is also a lame excuse for not trying harder. Such information is available on questions such as energy resource development; its wider dissemination to the public must be regarded as a task deserving the highest priority. This book is a modest offering toward that goal.

The approach taken in this study of the tar sands is an old-fashioned one known as political economy. It is quite impossible to separate politics and economics in analyzing resource development if one wishes to see the world as those who make the key decisions see it. Particular emphasis is given in the following pages to the nature of foreign corporate power in Canadian society and its close ties with those who govern the country. Much has been written on this subject by others far better qualified than myself, and I do not pretend that any new information contained in these pages will do much more than confirm some of our worst fears. Perhaps it will serve to disabuse us of the complacent belief that, in spite of its pervasive economic presence, the American corporation is not involved directly in Canada's political life. Contrary to that comfortable myth, scrutiny of the politics of Syncrude reveals that the multinational companies which own so much of our economic wealth are also highly purposeful, ruthless institutions engaged in what can only be called a struggle for power in Canada. Those who believe that we can isolate our political system from the effects of the steadily growing takeover of the Canadian economy by foreign corporations are either preaching deception or living in a fool's paradise. Monopolistic economic power conveys social and political power; there is no way in which we can quarantine our political and cultural independence of action while surrendering our economic sovereignty.

2　A Mess of Promises: Into the Syncrude Era

"It's never so bad it can't get worse."

Norwegian maxim

They danced in the streets the night Peter Lougheed ushered Canada into the Syncrude era.

Speaking on prime time provincial television on the evening of September 18, 1973, Alberta's Conservative premier announced that the giant billion dollar Syncrude project, ten years on the planning boards, was "go." It was, Lougheed affirmed, "an historic night." Months of "long, hard and tough" negotiations between the Alberta government and the Syncrude consortium of four U.S.-controlled oil companies had finally culminated in the group's decision to proceed "on the government's terms." Syncrude's complex for mining, extracting and upgrading the sticky tar sands to a valuable synthetic crude oil would create thousands of meaningful jobs for Albertans and represented the tapping of one of the world's largest reservoirs of unused energy—"almost half of the world's conventional oil supplies, a vital backup supply," as Lougheed measured it. The premier hinted at several similar oil sands projects in the planning stages and noted that, in view of the enormous size of the reserve, "obviously the market has to be essentially an export market, in the United States, possibly in Japan." Much of the speech, which began with a long eulogy on the contribution of the oil industry to Alberta's prosperity, was a justification of the government's "hard bargain" with Syncrude and the resultant "good deal for the people of Alberta." First, the government had insisted on receiving, in place of the usual royalty on gross production, a fifty percent share of

Syncrude's profits (here the message "fifty percent of all profits" flashed across the television screen). Second, the government was creating the Alberta Energy Company to act as a vehicle for public participation in the project: the company would own eighty percent of the pipeline facilities, fifty percent of the power facilities fueling the plant and would have a twenty percent option in the Syncrude plant itself. In all, Lougheed promised, Alberta would receive a billion dollar return from the project.

The premier spoke against the background of a steadily worsening feud between his government and the federal Liberal administration of Pierre Trudeau, a feud which had taken a recent turn for the worse when the federal government imposed a forty cent per barrel export tax on crude oil shipments to the U.S. Lougheed spoke darkly of a federal "power play to try to control the oil and gas resources of Alberta . . . . We intend to fight back, we have no other choice." Then he proceeded to dump the entire Syncrude deal into the laps of the "feds": the agreement was conditional on Syncrude winning some important tax and pricing concessions from the federal government. "Yes, the project could still be killed by the federal government. I'm confident they wouldn't do that." If they did, the effect would be felt not only in lost jobs, "but oil sands development might be set back permanently, because there are alternatives—the Colorado oil shales, nuclear energy—and of course Canadian crude oil backup supply would be weakened considerably." In brief, Lougheed was equating his deal with Syncrude with the national interest and daring Ottawa to do its worst.

Full marks go to the premier's image makers; it was a very effective job of salesmanship. "The television show was very slick," wrote one critic, "the best political broadcast yet seen in Canada." Others thought it a little overdone, resembling, with all the paraphernalia of electronic presidential politics, a page lifted from a Ronald Reagan speech. Still, as the president of Syncrude informed the participants in the consortium three days after the speech:

> "Syncrude's decision to proceed has sparked reams of news media comment. As expected, Premier Lougheed and his political associates have taken every advantage of the situation for political promotion—and not without results. The provincial government's violent stand against Ottawa on the oil export tax question coupled with the Syncrude announcement to proceed has made Lougheed a hero in the eyes of the "man in the street." Any criticism made to date by political opponents has

been meaningless and ineffective . . . . Premier Lougheed called today and personally congratulated and thanked Syncrude on the manner in which we handled the confidentiality of the negotiations with the government, and the manner in which Syncrude handled its end of the publicity with regard to the announcement to proceed."

As this letter points out, there was little immediate negative criticism of the deal and much lavish praise. The Toronto *Star* headlined its story, "Alberta Taking Piece of the Action," and the Edmonton *Journal* suggested in an editorial that it was "the kind of good news all Albertans can share." The Calgary *Albertan* and the same city's *Herald* were equally enthusiastic. Conservative M.P. Alvin Hamilton described the deal as "economic nationalism at its finest," and the head of the Canadian Petroleum Association, John Poyen, added that the province had negotiated "a pretty tough deal": "It looks like a meeting of the minds that can be very beneficial to Albertans." The oil companies backing the project—Imperial Oil, Gulf Oil, Atlantic Richfield and Cities Service—also seemed to think it was a fair deal, but there was a little confusion as to whose energy crisis Syncrude's oil was going to alleviate. Imperial's president, J. A. Armstrong, said the project "would represent the largest single investment made thus far in assuring continuity of Canadian petroleum energy supply," while Jerry McAfee, his counterpart at Gulf, thought the tar sands offered "the best prospects for maintaining Canada's important oil export trade with the United States." And up in Fort McMurray, one report even had a few citizens dancing in the streets after the Lougheed broadcast. Whether the dancers were of native or Chamber of Commerce extraction was not made clear.

Inevitably, there were one or two spoilers. The leader of the provincial New Democratic Party, Grant Notley, criticized the government for taking such a small share of Syncrude and wondered why the oil sands should not be developed by Canada through a crown corporation. Dr. Bruce Wilkinson, chairman of the Department of Economics at the University of Alberta and one of Canada's most respected economists, did a few calculations and came up with some intriguing results: at worst, the "fifty percent of all profits" scheme might amount to a royalty of less than one percent; at best, about seven percent. Either figure would be one of the lowest in the world. Wilkinson questioned whether the province would ever receive the promised billion dollars from the project. More likely, he suggested, the public's return would come to something between

$300 and $500 million over the project's life, less than it would end up investing in the venture and its related infrastructure. Other sceptics cautioned that profit sharing arrangements give multinational companies a decided incentive to transfer profits away from the production end of a venture such as Syncrude and into their downstream operations. The four oil companies would not be interested in seeing Syncrude itself make profits. Did Alberta have the know-how to prevent itself from being fleeced by accountants sitting behind desks in New York? What was a "profit" anyway? What were the companies being allowed to write off before the calculation of profits? One former Social Credit cabinet minister wondered whether "we might find ourselves receiving fifty percent of nothing."

Probably the most discordant note injected into the convivial atmosphere surrounding the Syncrude event originated with a rather unlikely source—the government itself. In a move timed to coincide with Lougheed's "go" announcement, the national chairman of the Committee for an Independent Canada, Mel Hurtig, made public a confidential report on the oil sands which had been drawn up for the provincial cabinet in mid 1972 by a large group of Alberta's senior civil servants, representing some twenty government departments. Describing the leaked eighty-page document as "one of the best statements in principle ever produced in Canada about natural resource development," Hurtig (who had received the report anonymously in the mail) charged that the spirit and terms of the Syncrude agreement stood in sharp contrast to the strongly pro-Canadian thrust of the recommendations of Alberta's civil servants. The government appeared to be thumbing its nose at its own top advisors. The civil servants' report did not bear directly on the Syncrude negotiations, but from its discussion of the ways in which the multinational corporations would oppose what it defined as the "primary objective"—development of the oil sands to meet the growing socio-economic needs of Albertans and Canadians—one could infer a good deal about the kinds of sentiments circulating within the bureaucracy of the government. The civil servants had proposed a reversal of what they saw as "the historical trend of ever increasing foreign control of nonrenewable resource development in Canada." More than two years after its drafting, the Conservation and Utilization Committee's "Athabasca Tar Sands Development Strategy" remains essential reading for those who think that Canadian governments have lacked good advice in their approach to resource issues. On the contrary, as we shall see, in the case of the oil sands an alternative to dependence and foreign corporate control had been spotted—and ignored.

Did it matter? What wider significance did Lougheed's September 18 speech hold for Canadians? Why should Canadians, particularly those living outside Alberta, care about whose advice the government took, or about the nature of Syncrude, or the terms of complex government-corporate agreements, or even about the long-run fate of the oil sands? What relevance had these unexploited and largely unknown resources lying beneath the muskeg of northern Alberta for Canadians living in Moncton, Prince Rupert or Kingston? Why, to put the matter in a nutshell, should the average, working Canadian bother about what Peter Lougheed planned to do with the oil sands?

The tar sands concern us all; every Canadian, and every Canadian's children and their children, have a direct stake in the future development of our energy resources. They should concern us because of the vast potential of the oil sands for delivering energy to a Canada which has profligately exported so much of its conventional energy reserves that a supposedly self-sufficient nation in oil and gas is facing serious shortages in this decade; because of the intense pressures suddenly being exerted by foreign interests to rapidly develop the oil sands, with scant regard for the benefits to Canadians; because of the traditional ways in which Canadian governments have foregone long-term benefits for short-term expediency and virtually given away our country's raw wealth; because an excessive concentration on huge export-oriented resource projects has serious effects in other important areas of the economy; because Canadians have been through other ''boom-bust'' cycles launched on someone's grandiose visions and know from sad experience that history often repeats itself; because of the unequal, unfair, unjust ways in which the material rewards of resource development have been distributed in the past, with a few growing fat from the quick sale of what belonged to all and the rest living in ''next year country'' on a mess of promises; because of the intensifying hold of the new monopolists, the great energy conglomerates answerable to no government or nation, over the economic and political possibilities of Canada. To leave the oil sands under their ownership was like issuing them an invitation to take out an insurance policy against the possibility that another generation of Canadians, less short-sighted than our own, might decide to choose their own road.

If Canadians do not take an interest in the tar sands, others certainly will. Shortly after the Syncrude announcement several other large companies began to queue up for position. Government ministers issued a call for "crash programs" of development with as many as twenty Syncrude-sized plants in place by century's end. Even more extravagant schemes have been proposed in the United States and taken up by gullible Canadian politicians. International energy developments since 1973 have added to the pressure emanating from without on Canada's untapped resources of fuel. New power centres and aspiring monopolists, challengers to the traditional dominance of the majors of the petroleum industry, have changed the rules of the game in world oil, making the tar sands and other alternate forms of energy seem a less prohibitive proposition. The largest international energy companies have embarked on diversification strategies, searching for new reserves, new surplus capacity and increased political bargaining power. What fuels this policy is not the spectre of imminent oil scarcity in the world, but a struggle for wealth and power.

As pressures for development mounted, many of them originating outside Canada, how would the governments of Alberta and Canada react? The strategy of the oil industry is to try to persuade the public sector to underwrite as much of the cost and risk involved in developing new, expensive energy as it can get away with. And what it has been able to get away with in the past has been a very great deal. The new oil sands play was opening up just as Canadians were starting to ask some urgent questions about the management of their energy resources. If the past record of resource regulation and development and the traditional vacuum of public policy persisted, then Canadians had every reason to be alarmed.

How was it, many Canadians were beginning to wonder, that the country was now apparently facing energy shortages? Had we not been assured by cabinet ministers that we had literally centuries of energy reserves stored in the ground, and that we must sell them abroad before they became obsolete? Were we not so rich in unused energy that our real problem lay in finding buyers for the surplus? Why were we now hearing other voices tell us that Prime Minister Trudeau must have got his facts wrong: Canada might be self-sufficient in oil until 1980, but certainly not after 1980! Why were we facing the paradox outlined by Dr. F. K. North, a geologist: "We may have resources for 100 years, but by the late 1970s and until the late 1980s, we'll be exceedingly cold!" "How is it—and why is it," we are entitled to wonder with writer John Aitken, "that we have come to this crisis in energy when just a few years ago our main concern was that the Americans might not buy our oil as rapidly as we could pump it out of the ground?'' How indeed?

The oil companies have their own explanation and solution for this scandalous state of affairs. It was necessary, they inform us, to export so much Canadian oil and gas to the United States because the Canadian market was too small to achieve the necessary ''economies of scale'' required to develop and move the resources. Now it will be necessary to develop new energy reserves—in the Arctic, oil sands and offshore—to replace what has been sold. But the difference is that these new reserves will be much more expensive to develop. Thus, the oil companies say, they must have a considerably higher rate of return on our remaining reserves of conventional oil and gas in order to pay for their higher exploration and development costs. Moreover, the economies of scale problem still demands that a heavy ''export component'' be built into the development plans for these new energy reserves. The market must continue to be a continental market; otherwise, Canadians will never be able to develop these new reserves which are needed to overcome the shortages, which are in turn the consequence of the way in which the western Canadian reserves were depleted. In short, continental solutions to continental problems. In order to replace our mismanaged resources, the best and cheapest energy Canada ever had, we are now being offered crash programs for the oil sands, east coast superports, the Mackenzie Valley gas pipeline and the implied threat that if the oil companies are not permitted high profits and given carte blanche to make their export arrangements with their U.S. parents, then the energy will be left in the ground and Canadians, presumably, left to freeze in the dark.

This circular reasoning, which appears to doom Canada to a perpetual cycle of shortages, high prices and foreign ownership, is actually a specious, self-serving piece of corporate rationalization—an apologia for economic imperialism. As such, it deserves an answer as contemptuous as the spirit in which it is offered. Yet among those who design provincial and federal energy policies the ''deadly logic'' of the oil companies seems to be taken at face value. Thus we have heard Peter Lougheed denounce suggestions that Alberta ''keep its oil in storage for tomorrow in order to serve a slow-growing eastern Canadian market, or have its oil available for some kind of emergency,'' yet not hesitate to threaten ''conservation'' (i.e., interrupt oil deliveries to eastern Canada) in protest against Ottawa's policies. And we have had a federal minister of Energy, Mines and Resources promise ''a new round of resource exploration'' and sound for all the world like the chairman of Imperial Oil. So long as Canadian governments continue to think like foreign-controlled oil companies, any formal continental energy treaty is clearly redundant. All they have to do to get at our energy is get at our politicians.

In view of the oil industry's game plan and the demonstrated willingness of Canadian governments to play along, Canadians should now be seriously worried about the status of their future energy reserves—the oil sands, the reserves of the Mackenzie delta and high Arctic, the offshore reserves, the uranium and coal deposits. For these sources of future fuels are already in the hands of many of the same corporations who control Canada's remaining conventional oil and gas—along with much of the coal and uranium of the United States, the leased portions of the Colorado oil shales, the oil and gas of the north slope of Alaska, oil deposits in the North Sea, Latin America, Africa and southeast Asia, as well as major, albeit vulnerable, holdings in the Middle East. How and when the highly inbred energy industry decides to bring one of these resources into production or cut back on another has little to do with the particular needs or priorities of the nations in which the affiliates operate. Rather, such decisions spring from the overall economic and political strategy of the global corporation, ever in search of flexible advantageous bargaining postures, new reserve strength and more profitable investment opportunities.

In negotiating with the affiliates of these giant corporations governments are often at a decided disadvantage, for the highly centralized companies can pit provinces, countries, regions, even different fuels, against each other in order to win the terms they seek. And where a government is isolated and sees no alternative to the multinational corporation and its priorities, then that government will have little power to dispose of its resources as it thinks best. It is this situation which accounts for the rise of international organizations like the Organization of Petroleum Exporting Countries (OPEC), founded to counter the bargaining power of the world oil cartel. And it is precisely the same situation which has placed Canada's future energy supplies in question and left our governments open to corporate blackmail. Governments who play poker with companies like Syncrude are playing against a stacked deck.

If such considerations do not seem particularly troubling, recall that they have consequences for our heirs as well as ourselves. How we develop our new energy resources; who develops them; for whom; and at what cost—these crucial questions bear on the opportunities for other than our own generation. Each generation is to some degree the prisoner of its predecessor's choices. Will our children one day find themselves tied into fixed positions, paying excessive prices for domestic energy or facing the threat of embargoes because of decisions taken at the opening of the Syncrude era by those whose eyes were filled only with the visions of unbounded growth and a place in the North American sun? "Build more pipelines to export more oil and gas now?" asks Eric Kierans, former federal cabinet minister and now a professor of economics at McGill University. "Tell it to your children! Or better, ask them what they think of a Canada that gives away its gifts of energy for a mess of promises about fast-breeder reactors and solar system breakthroughs." Kierans is speaking here of the proposed Mackenzie Valley gas pipeline, but his point applies with equal force to the oil sands. As a model for developing the new resources of the nation, Peter Lougheed's deal with Syncrude would mortgage the future of Canada.

An alternative, strongly documented and logically argued, lay in the recommendations of the government's own civil servants: recommendations which, as they themselves noted, were characterized by their pro-Canadian flavour. The report was a plea for change, for reversing the pattern of foreign control of Canadian resources and resisting the pressures of the multinational corporation. There were, according to the civil servants, two diametrically opposed courses of action that were open to Albertans and other Canadians. In the first place, they could "continue the policies of the conventional crude oil developments creating tremendous and unregulated growth and developments resulting in short term benefits accruing to the Province as well as the long term costs arising from exported energy, technology, job opportunities and environmental damages, in addition to the depletion of non-renewable resource [sic]." Or, "we can regulate the orderly growth and development of the bituminous tar sands for the ultimate benefit of Alberta and Canada in order that Canadian technology will be expanded, Albertans will find beneficial and satisfying employment within its diversified economy, and our environment will be protected and enhanced for future use." As the civil servants assessed it, there was really no choice here: when the costs and benefits were weighed "it becomes apparent that the latter course of action is imperative."

The frank and unambiguous nationalism of the Conservation and Utilization Committee flowed from a fear that this last and greatest of Alberta's nonrenewable resources would be developed in the same manner as the conventional energy reserves of the post-Leduc era.

Their biting description of that period (and who was in a better position to know), the "tremendous and unregulated growth," the "short term benefits" and "long term costs" arising from exported energy, technology and jobs, was an interesting confirmation of what energy critics had been arguing for years. Pointing to the fact that the pressure on the oil sands was largely emanating from outside the country, the civil servants urged the Lougheed cabinet to resist the temptation to allow foreign energy demands to dictate the terms of development. Instead, the scale, timing and staging of oil sands exploitation should be determined "by deliberate policy decisions by Albertans designed to ensure that the development complements and supplements the overall development requirements of Albertans and Canadians. The policy decisions should be guided primarily by the perceived benefits that will accrue to Albertans and Canadians. Only after Albertan and Canadian policy parameters have been fulfilled should foreign constraints become operative. In short, Canadian policy parameters should take precedence over all other factors."

These objectives, the civil servants warned the cabinet, would be "vigorously opposed" by the multinational corporation since they would diminish its control and minimize its profits. The oil companies were interested in rapid development on their own terms and conditions: they would tend to import equipment, engineers, managers and staff and in turn export the synthetic crude in an unprocessed form. Foreign corporations controlled the leases, crucial information, the extraction and processing technology and the capital, and they were uninterested in questions about Canadian economic sovereignty. "The attitude and expectations of the multinational corporations proposing the development is premised on investment opportunities throughout the world, the size of their investment and the ultimate rate of return on investment. In order that they may maximize their profits they will tend to externalize as many of the costs arising from the project as can legitimately be done. Since the environmental costs of this development are extremely high and since the current technology and economics of extraction are still in their operational infancy, the tendency will be for the corporate structures to externalize these costs for society to absorb." Alberta would have to "stand firm in the conviction that the tar sands make up approximately one-third of the known world petroleum reserves," and the government would have to engage in a program of strict corporate surveillance. The civil servants did not take the next crucial step and call for public development of the oil sands—and it would be unrealistic to have expected them to do so—yet they were firm that their entire policy was "totally dependent on Canadian economic

28

participation and control of the consequences of the future development.'' Without this, any "pro-Canadian" strategy would be a sham. Here in broad outline was an alternative to Syncrude. Here was an opening for the move, the break away from the historical pattern of foreign control of Canadian resources.

As mentioned the civil servants did not take their argument to its logical conclusion. Public ownership and control of oil sands development, rather than a passive, minority public "participation" in such ownership, would appear to be the most effective method for implementing the pro-Canadian policy they proposed. Yet the recommendations of the Conservation and Utilization Committee, if fully and aggressively backed by government, would still take Canada a substantial step beyond her present dependent status. Moreover, the committee's cautious suggestions for containing and mitigating the potentially severe social, economic and ecological effects of rapid development of the oil sands stand as a stiff warning against the mindless exploitation of nonrenewable resources, men and the environment, of which the world has seen far too many examples.

The "imperative" policy choice argued by the Alberta civil servants has been consistently ignored, violated and abused by Canada's politicians. Far from seizing the occasion to reverse "the historical trend of foreign control of Canadian resources," the Lougheed government caved in to the pressure of four American-controlled oil companies during negotiations in the summer of 1973, then strongly backed the consortium in its dealings with the federal government. In virtually every case, as we shall see, the development conditions and terms of the oil companies have been met by Canada's public leaders. Indeed, most of the heavy costs of the project are being carried by the taxpayers of Canada, Alberta and Ontario through minority equity holdings, loans, foregone royalties and taxes and the burden of infrastructure expenses. The oil companies were able to shift much of their costs and risks to the public sector following the sudden withdrawal of one of the four original members of the Syncrude group in December 1974. For two tense months the big tar sands project made headlines in Canada as the remaining corporations laid down a further list of conditions for keeping the venture alive. The outcome of the Syncrude poker game was predictable enough: Syncrude was rescued on terms dictated by the private companies in an extraordinary ultimatum of mid January 1975. If this was an indication of the nature of the Syncrude era, Albertans and all other Canadians could be forgiven for wondering what the dancing had been about.

3 "Welcome to Our Oil Sands . . ."

"We are anxious to have the tar sands developed as soon as it is economically sound to do so, and no private party of sufficient financial strength is coming forward to test the economics of the matter. It looks as though this is the sort of task for a government to undertake, and since the government of Alberta controls the tar sands and would benefit by profitable development, the government of Alberta is indicated as the logical one to build and operate the plant."

Dr. Karl A. Clark
January 1945

"The government of Canada does not trade oil for gas or oil for bananas or gas for peanuts and it would be very wrong to do so. The only thing that we can do in the national interest is to sell the surplus . . ."

Federal Energy Minister J. J. Greene
October 1970

The early road to Syncrude was paved with good inventions—all of them Canadian.

We can trace the sell-out of the tar sands back to a remarkable week in the autumn of 1951 when the Social Credit government of Alberta, headed by Premier Ernest Manning, invited the cream of the international mining and petroleum industries to take over the future of one of the world's greatest untouched resources. Looking back with the advantage of hindsight, the occasion now seems full of irony and symbolism. At the first Oil Sands Conference, held on the University of Alberta campus in Edmonton over the week of September 10-14, 1951, four decades of frustrating, lonely pioneering work in the wilderness of northeast Alberta by scientists of the Dominion and provincial governments culminated in the decision to place the vast bituminous sands under the monopoly control of the globe's biggest resource extraction companies and to turn forty-odd years of publicly financed research and technical know-how over to their profitable advantage.

The invitation to the tar sands could hardly have been more effusive. "Through the years," one provincial cabinet minister told the assembled international businessmen, "the words 'welcome' and 'Athabasca oil sands' have been very closely associated." The speaker recalled how Peter Pond and Alexander Mackenzie, the first white men to set eyes on the tar-saturated banks of the Athabasca River, had been welcomed by the Indian tribes of the area. Poor Mackenzie had even parted with a precious flask of whiskey to purchase some of the molasses-like tar for canoe repairs. "It is indeed a far cry," continued the Social Credit minister, "from Peter Pond and the oozing 'tar' made to run by a powerful sun, to the modern pilot plant; a century separates the primitive gumming of canoes by Sir Alexander Mackenzie and the modern uses made of petroleum. But the warmth of the welcome accorded to you by the people of Alberta tonight is just as radiant, just as sincere, just as captivating as that accorded the early explorers by the primitive red men. May the words 'welcome' and 'oil sands' ever be closely associated!"

In keeping with these sentiments, the Manning government had arranged for a detailed presentation to the world's oil and mining leaders of virtually every bit of technical information then available on the geology of the oil sands, the location of the richest and most accessible deposits, the various extraction techniques developed by government researchers for recovering the raw bitumen from the oil sands, the methods for upgrading the thick, heavy oil to valuable synthetic fuels, the costs and profits likely to be realized from a commercial operation, and—what else?—an analysis of Canadian tax laws. Social Credit had even gone to the trouble of running a pilot plant in the late 1940s to prove that the tar sands could be exploited commercially by free enterprise. For good measure and as an added incentive, an oil sands leasing policy of uniquely generous dimensions, even for postwar Canada, was announced to the conference on September 11, 1951, by Nathan Tanner, Alberta's minister of Mines and Minerals. Alberta, said Tanner, was "opposed to monopoly of any kind, whether it be the government, a company or a group of companies. We feel that only through individual enterprise, where we have good wholesome competition, can development go forward as it should go forward, and we are determined to see that that same competition, individual free enterprise as you refer to it, is carried on and our natural resources developed under that program." Alberta would reward those who invested in her resources, yet she expected a fair share of the benefits: "We would like you to know that we are going to try to be at the end of the rainbow to get our share at the same time."

Alberta's government, said Tanner, "is desirous of doing all that is reasonably possible to encourage the orderly development of the enormous oil sands' deposits in the interest of the people of the province and of Canada as a whole, and, further, to the security of this continent." It had been decided "to encourage immediate development to meet the ever increasing demand for petroleum products and to offset the effect of the uncertainty of supply elsewhere in the world." Prospecting permits and leases would be issued on a first come, first served basis in blocks up to 50,000 acres, one block to an applicant. Permits would be granted for one year, renewable for two years, for a fee of five cents an acre the first year, ten cents an acre for the second, and twenty-five cents an acre for the third. Provided the permit holder had furnished evidence of exploration, he would then have the exclusive right to acquire leasing privileges in the bituminous sands. The lease would grant the exclusive right to develop the oil sands for a twenty-one-year period, renewable for additional twenty-one-year terms, for an annual fee of one dollar per acre (this was later reduced to twenty-five cents per acre). Royalties charged on any development, Tanner promised, would not exceed ten percent of the value of the raw bitumen extracted. Finally, and perhaps most important, there would be no further crown development of the tar sands.

Here was a "welcome" indeed. Having first attacked the idea of monopoly, Social Credit had then devised a policy which would in the course of a few years introduce a complete monopoly over the huge energy reserves lying in the tar sands. At first this power would be exercised in the decision simply not to develop the tar sands, to tuck them away as an insurance policy for the future. Later the insurance policy would be cashed in and the power of monopoly would then be exercised as the demand to develop the tar sands on the terms of the lease holders. What Nathan Tanner planted as a seed at the 1951 Oil Sands Conference would ripen into the politics of Syncrude more than two decades later.

Not surprisingly, in light of the immense potential wealth they are known to hold, Canada's bituminous sands have had a colourful and often controversial history. The existence of the great Athabasca deposit has been known for many years, and it has naturally attracted its share of explorers, fortune hunters and genuine developers. In the closing years of the eighteenth century the fierce fur trade rivalry between those early conglomerates, the Hudson's Bay Company and the North West Company, saw the first of the adventurous Nor'-westers probing deep into the Athabasca country and beyond to the Mackenzie. They came by canoe from the headwaters of the Churchill across Methy portage and down the Clearwater River. About 250 miles northeast of Edmonton the Clearwater meets the Athabasca, flowing east and north toward Lake Athabasca, and there on the valley floor at the junction of the two rivers the Nor'wester Peter Pond founded the "Fort of the Forks," today the site of Fort McMurray, in the year 1778.

Pond paddled due north down the Athabasca and saw the strange layers of tar sands—later to be named the McMurray Formation—cut into the steep river banks. Wintering south of Lake Athabasca, Pond learned that the Indians used the tar to caulk their canoes and for medicinal purposes. A few years later his better known successor in the north country, Alexander Mackenzie, followed Pond's route down the Athabasca and marvelled at the oil-impregnated earth near the spot where today stands the Great Canadian Oil Sands complex. "At about twenty-four miles from the Fork," recalled Mackenzie, "are some bitumenous [sic] fountains, into which a pole of twenty feet long may be inserted without the least resistance. In its heated state it emits a smell like that of sea coal. The banks of the river, which are there very elevated, discover veins of the same bitumenous [sic] quality." It sounded temptingly simple.

What Pond and Mackenzie—and later David Thompson and Sir John Franklin—stumbled upon in their trips north was but a tiny exposed fraction of one of the world's richest reservoirs of untapped oil. What are the "tar sands"; where did they come from; what distinguishes them from ordinary crude oil; and just how big are these deposits which are now attracting so much interest?

The tar sands, sometimes called "bituminous sands" or "oil sands," consist of a varying mixture of sand, clay, minerals, water and crude bitumen. Bitumen is found around the grains of sand, separated by a thin film of water. In some areas north of Fort McMurray the tar sands can be found exposed or lying very close to the surface; usually, however, the bitumen is buried beneath an overburden of soils and clays up to 2,000 feet in depth. Where a deposit is particularly rich the layer of tar sands may be 150 feet or more thick, the bitumen making up about twelve percent by weight of the sands. The sand looks and feels rather like asphalt paving material, and in fact one of its earliest uses was in the paving of streets in McMurray, Edmonton and Jasper Park. How and where the oil sands originated is still something of an unsolved riddle to scientists. Some geologists think the bitumen used to be a lighter oil which migrated from other strata to its present reservoir and then bituminized or became tar-like; but others believe it is really a virgin oil, a proto-petroleum, which is in the early stages of developing into a lighter and more familiar petroleum.

The basic raw material in question, the so-called "tar," is bitumen, a very heavy, black viscous crude oil which in a liquid state resembles the slowest and thickest molasses. Cool some in a glass, turn the glass upside down and the stuff will not run. The oil is found in the bituminous sands of the huge Athabasca deposit, and it is also found (or closely resembles the oil found) in deeply-buried heavy oil deposits elsewhere in Alberta. Crude bitumen is not an especially attractive hydrocarbon: for one thing, it is too heavy and viscous to transport by pipeline; for another, it is deficient in hydrogen, that is, it has a lower ratio of hydrogen to carbon than conventional oils; and for another, it is high in sulphur content. But crude bitumen can be processed and upgraded to a valuable and highly flexible synthetic product when it is broken down, coked, desulphurized and hydrogen is added. This yields a high quality, semi-refined blend of naphtha and light and heavy gas oils which is exceptionally sulphur-free. The oil companies call this straw coloured blend of liquids "synthetic crude oil," but this incorrectly implies that the product is merely a substitute for conventional crude. It is not, and the oil industry knows it is not. Rather, it is of a higher quality and apparently has many more potential uses, particularly as a petrochemical feedstock, than ordinary crude. What prevents that potential from being realized is that the product is being integrated into the refining, marketing and pricing structure of the oil companies. The lease holders define the use, hence we have "synthetic crude."

By any reckoning, a lot of bitumen is lying around northern Alberta. The full extent of Alberta's four major oil sands and heavy oil deposits is not completely known, but one recent estimate suggests that the total amount of crude bitumen in place may be above 900 billion barrels and that up to 250 billion barrels of synthetic fuels are ultimately recoverable. The largest and best known deposit, Athabasca, is thought to hold reserves of above 600 billion barrels of bitumen in place. The Cold Lake heavy oil deposits near the Saskatchewan border, where Imperial Oil has a major pilot program under way and where the Japanese are becoming involved, are figured to hold another 164 billion barrels. The Wabasca deposit is not well known and leasing is still going on, but it may contain some 54 billion barrels of bitumen in place. And at Peace River, where Shell is the major lease holder and is operating a pilot project, another 50 billion barrels of bitumen in place have been estimated.

When one considers that Canada's remaining reserves of conventional crude oil are reckoned at only eight or nine billion barrels, these figures seem absolutely staggering. But they are also highly misleading. To recover and develop even a small fraction of the total oil will be extremely expensive, time-consuming and will involve massive environmental disturbance. Bitumen in the ground is not gas in the tank by a long shot. A great deal of inflated and overly zealous rhetoric about the size of oil sands reserves has already inspired an irresponsible attitude towards the development process. However, it is now clear that we must scale down our expectations considerably and begin to think of the tar sands in terms other than their sheer size. Properly and cautiously developed, the sands could make an important contribution to Canadian economic and social development. But, in view of the costs and sacrifices we will have to make to develop substantial amounts of the tempting "theoretical," "ultimate," or "recoverable" reserves, Canadians may be better off if most of the tar sands remains in the ground for a long while yet.

In any event, technologies do not yet exist for exploiting most of the tar sands and heavy oil deposits. The existing commercial technology, as we shall see, was developed slowly and through a long, cumulative process, and in many ways it is still very imperfect, particularly from an ecological standpoint. The bitumen does not flow naturally and it cannot be pumped; hence, it must either be made to flow, or the sands themselves must be mined and the oil extracted in a subsequent operation. The latter technique is the one being used or proposed by GCOS, Syncrude, Shell, Petrofina and Home on their leases north of McMurray along the Athabasca River. The approach involves open-pit mining of the sands and then hot water extraction of the bitumen, followed by upgrading to synthetic fuel. But this can only be done where the tar sands lie close to the surface: only deposits lying less than 150 to 200 feet beneath the muskeg and overburden of soils are presently deemed to be recoverable through the surface mining technique. Less than five percent of the total oil sands of Alberta falls into this category, and all of it is concentrated in one strip of the Athabasca deposit lying due north of Fort McMurray along either side of the Athabasca River. This small zone of roughly 480,000 acres, seventy-five miles long and twenty-five miles wide, contains the prime leases and it is this thin ribbon of Canadian territory which is the focal point of today's oil sands play. Here it has been estimated that about 26 billion barrels of synthetic oil could be recovered by digging for it.[1]

Where the tar sands are too deeply buried to be mined, the bitumen can only be recovered through some kind of *in situ* (literally, "in place") technology aimed at heating, dissolving or emulsifying the heavy, viscous oil and getting it to flow to production wells. There are no *in situ* plants in commercial production, but Imperial, Shell, Texaco, Amoco (Standard of Indiana) and other companies have all had major pilot projects in one or other of the major deposits. Research into commercially viable and environmentally sound *in situ* processes is being accelerated—emphasis is currently being given to thermal stimulation techniques through the injection of steam into the reservoir for an extended period of time—but any major production from the ninety-five percent of the oil sands which are too deeply buried to be mined is unlikely before the mid 1980s. Consequently, and because the costly surface mining plants can only be built slowly and one at a time, Canadians will be lucky to be realizing much more than a bare half-million barrels of synthetic fuel per day from the tar sands by 1985.

The effort to force the stubborn bituminous sands to yield up their riches has been going on a long time. The earliest promoters, lured to the Athabasca forests by stories of the fabulous oil deposits, trekked into the remote and isolated bush with their patented inventions, one and all determined to "solve the riddle of the tar sands." Some, like the aristocratic German, Count Alfred von Hammerstein—eccentric and resplendent in buckskins, kerchief and Teddy Roosevelt hat—tried to pump out the bitumen with conventional oil rigs and predictably failed. Others tried to set the tar sands on fire or to have micro-organisms feed on them. They blasted it, hacked at it, mixed it, stirred it and finally damned it. The quest for the elusive, magic solution has not ended yet: every few months the newspapers headline another press release by someone claiming, at long last, to have the key. One imaginative tar sands enthusiast dreamed up the idea in 1957 (while gazing at a flaming sunset in the oil fields of Saudi Arabia) of using the power of an underground nuclear blast to get the bitumen flowing. For a time the idea was touted by Richfield Oil Company, one of the principals in the Syncrude project, but the scheme ran afoul of Canada's disarmament image about 1960. From time to time someone revives this suggestion of using the ultimate weapon on the bitumen but, fortunately, it remains very much in the realm of the hypothetical.

The early promoters and developers came to the Athabasca tar sands and faced the worst possible odds, and they invariably departed in frustration and bitterness. Fort McMurray grew up and survived almost in spite of the huge resource on which it sat. Founded by Henry John Moberly in 1870 as a steamboat terminus for the Hudson's Bay Company, McMurray later became a staging area for communications between Alberta and the Northwest Territories. After 1916 freight came up from Edmonton on the Northern Alberta Railway, then was transported to barges moving down the Athabasca and Mackenzie rivers all the way to the Arctic coast. Later, the town helped supply Uranium City on Lake Athabasca during the years of the uranium boom. Came the uranium bust and the development of alternative transportation routes to the north, McMurray, in the words of Earle Gray's *The Great Canadian Oil Patch,* ''shrank into a pocket of poverty in the midst of the affluence of oil-rich Alberta. . . . The people of Fort McMurray no longer cared. They had long since lost all faith and hope in the plans of the oil men. They knew that literally beneath their feet lay the largest known deposit of oil in the world, and for all that they could tell it would be there forever.'' Even now, when growth is straining the town to the limit, one can find longtime residents of McMurray who worry that the boom will be brief and that the community is launching grandiose plans for the future on the shaky foundation of a few promises. What might happen, they wonder, if the promises are not made good and the bubble bursts? That is a good question and it reflects a shrewd appreciation of Fort McMurray's present dilemma. Having waited for the oil men for so long, the town now finds its future inseparable from investment policy decisions made thousands of miles away for reasons which have nothing to do with its own interests and needs. But to a degree, it has always been that way.

Why are the tar sands being developed only now? Why, for example, did large-scale extraction not begin after the Alberta government's elaborate demonstration in 1951 that it could be done—and for a decent return? Or, why not in the 1960s instead of the 1970s? These questions may seem academic, but they are not: the answers provide some important clues to the future, and the future of the tar sands should interest every Canadian.

If we look back over the history of the tar sands, we see that their development has been impeded by a tangle of technical, economic and political obstacles. In the first place, unique technologies had to be evolved to extract the bitumen from the sand and upgrade it to a clean, marketable product. Secondly, even if the technological problems had been surmounted, development on an economic scale would be very costly. Few developers could afford to risk the capital required for a major oil sands project. Third, the resource was remote and isolated, weather conditions were formidable, and communications and transportation to the Athabasca area were clearly inadequate to support heavy-scale development. These were the concrete realities which defeated the early developers and promoters; overcoming them would be difficult and time-consuming, though not impossible. But in addition to these hurdles, a fourth impediment to development has been decisive. Strong governmental and corporate pressures against development have played a crucial role in the recent history of the tar sands. At least since the Second World War, the decision whether or not to mine the tar sands has really been a matter of power politics, a matter of political rivalry for control of the future of this immense resource.

Canadian scientists and technologists were ultimately responsible for "solving the riddle of the tar sands." The basic technology for separating the thick, black oil from the sand was evolved, not invented, through a trial and error process over a period of some forty years. The "hot water flotation" method used today at GCOS, the method planned for all the surface mining operations, was worked out in the laboratories of the Alberta Research Council and in pilot plants in the oil sands from the 1920s to the 1950s. The man who led the Research Council's assault on the bituminous sands and whose name is usually associated with the hot water process was Dr. Karl A. Clark. Clark became interested in the tar sands in the early 1920s and began experiments in Edmonton, then operated a very small pilot plant on the Clearwater River near McMurray in 1929-30. In a 1951 publication Clark described how the stubborn sands could be made to yield up their oil. "The operation is a simple one. The oil sand is mixed and heated with water to a pulp containing twelve percent to fifteen percent of water. This pulp is then flooded in excess hot water. The oil separates from the sand and floats to the surface as a buoyant froth. The sand sinks. Fine mineral matter and some oil in fine fleck form becomes suspended in the plant water. The oil froth is skimmed from the surface of the plant water; sand tailings are removed by suitable mechanical means; and the load of sediment in the plant water is kept within bounds by some form of settling."

Clark was being overly modest; in fact, the process was not that simple. Hot water extraction (and other techniques) worked for others as well, but only the Research Council's method recovered over ninety percent of dirt-free oil.

In the late 1940s Clark helped build and operate a $500,000 pilot plant at Bitumount, fifty miles north of McMurray, for the Alberta government. The plant had its own mining operation, extraction plant, refinery, transportation system and tailings pond, and was in fact a much scaled down model of today's giant surface mining complexes. "The Bitumount plant performed well," wrote Clark; about ninety percent of the oil had been recovered from the processed sands. The way was open, Clark argued, for the study of "a complete sequence of operations, from mining to marketing." Such a study "would give the answer to a question that interests everybody, namely, whether bituminous sand development belongs to this generation or to a future one."

Clark's contribution was immense, but other government scientists also left their stamp on the tar sands. Until 1930, when the natural resources of Alberta were turned over to the province, the tar sands were under Dominion jurisdiction, and even after that date, as we shall see, Ottawa retained a strong interest in the resource. Dominion involvement, later to become an issue of bitter contention, began in 1882 when Dr. Robert Bell of the Geological and Natural History Survey of Canada examined the tar sands area, identified their geological age, and proposed that hot water extraction of the bitumen might be feasible and that a pipeline from Lake Athabasca to Hudson Bay should be built to transport the oil to foreign markets. Another Dominion geologist, R. G. McConnell, reported six years later that the tar sands "evidence an up-welling of petroleum unequalled in the world."

In 1913 the federal mines branch sent out a young engineer, Sidney C. Ells, to survey the economic possibilities of the tar sands. It was Ells who brought out the first big loads of the strange oil-soaked sands for study and who mapped the largely unknown deposit, and it was he who proposed the use of the sands for paving material. In 1915, just to prove his point, he brought sixty tons of the stuff out of the bush to Edmonton by horse team "in temperatures ranging from twenty to fifty below zero and without tents for men and horses." In Edmonton he paved some streets and later he did some paving with the bituminous sands in Jasper Park. Older residents of Fort McMurray remember Ells as an incredibly hardy lone wolf who devoted more than thirty years of his life to working in the oil sands under some of the harshest conditions imaginable. One longtime friend recalls meeting Ells one wintry morning, trudging toward McMurray along the frozen surface of the Athabasca. Ells was hungry—"Could use some grub"—he had exhausted his supplies and had been subsisting on nothing but rice during his week-long trek up the river in sub-zero weather. On another memorable day, this time in summer, the same friend encountered Ells drifting down the Clearwater River toward McMurray on a tiny makeshift raft. The young surveyor had lost everything in a spill in rapids and had been living on berries for several days.

There was another colourful tar sands pioneer, the first real entrepreneur to tackle the sands, who was active in these early years. Robert C. Fitzsimmons arrived in McMurray in 1922 "to investigate the possibilities of obtaining oil from the bituminous sand about which I had heard fabulous reports." Fitzsimmons bought out a New York outfit called Alcan Oil Company and set up his International Bitumen Company in 1927 at Bitumount. He first tried drilling for bitumen, then in 1930 developed his own imperfect version of the hot water extraction technique. Fitzsimmons produced several thousand barrels of unrefined bitumen in 1930 and 1931 which he later claimed "was excellent for paving, laying built-up roofs, processing into roof coatings, plastic gums, lap cement, caulking compounds, waterproofing, marine gum, fence post preserver, boat pitch, belt dressing, mineral rubber, and skin disease medicine." In fact, Fitzsimmons had great difficulty marketing his wonder products and also ran into obstacles while trying to raise enough money in Europe and the United States to construct a refinery. He later charged that his Bitumount operation had been deliberately sabotaged by an Alberta government intent on impeding large-scale development in the tar sands. In 1942 International Bitumen was taken over by Montreal financier Lloyd Champion and after a confusing series of corporate name changes emerged as Great Canadian Oil Sands. In 1953 a still bitter Fitzsimmons published an angry pamphlet entitled "The Truth About Alberta Tar Sands," in which he gave his interpretation of what had happened to his company:

"You may find it hard to believe that the tar sands were purposely kept out of production unless you understand that the Major Oil Companies must have oil reserves so that they know thirty to fifty years ahead where their next source of supply is coming from. Consequently they spend millions of dollars searching for new fields, but the tar sands was one source that they did not have to search for and they were determined to have that held in reserve until all oil fields recoverable from wells ran low, when they open up that area. Which, according to their plans, may be another twenty to thirty years.''

Under exclusive control of the international oil industry, Fitzsimmons argued, the oil sands would not be developed for another generation. Rather, the companies would sit on their leases, blocking production, and keeping the oil sands in trust as insurance against the future.

At about the same time that Bob Fitzsimmons' luck ran out at Bitumount a political storm broke out in Alberta over the wartime activities of the federal government in the oil sands. In 1929, a year before the Dominion turned over the tar sands and all other natural resources to the province, Sidney Ells had persuaded Denver oil man Max W. Ball to take an interest in the Athabasca sands. Ball eventually set up an operation, Abasand Oils Limited, near Fort McMurray, ran a pilot plant and in 1940 began extracting and processing some 400 tons of tar sands a day. But Abasand's extraction plant was destroyed by fire in November 1941 and did not resume operations until the summer of 1942. It was at that point that Mackenzie King's government, worried about its wartime oil supplies, moved into the tar sands. Probably at the urging of Ells, the Dominion had kept a foothold in the sands in 1930 when it reserved 2,000 acres of prime land, ostensibly for use in the paving of roads in the national parks. In June 1942 C. D. Howe, federal minister of Munitions and Supply, advised Alberta that Ottawa wished to explore the possibilities for a large-scale operation at the Abasand site. Under an agreement with the Consolidated Mining and Smelting Company, the Dominion took over the Abasand plant and also drilled a number of promising deposits in the area of Mildred and Ruth lakes. There a team of Dominion surveyors discovered a very rich ore body of bituminous sands—the same which would later be acquired by Sun Oil and which is currently mined by Great Canadian Oil Sands.

Social Credit, whose relations with the federal government were not the best, took a very dim view of Ottawa's new interest in the tar sands. The provincial government interpreted C. D. Howe's letter as a blatant attempt to use the excuse of national security to muscle in on a very valuable provincial resource. In 1944 the Manning government demanded a royal commission inquiry into Dominion activities in the tar sands. Charging a ''wanton plunder of provincial rights,'' the Alberta government accused the Liberals of trying to sabotage and discredit the possibility of large-scale production from the tar sands. In the words of an Alberta cabinet minister speaking in 1944, "shortly after the Dominion government took control of the property, strange things began to happen. Out of the north came an endless string of weird, incredulous stories of criminal incompetence, of scandalous waste of public funds and charges of sabotage were heard on every hand." Ottawa had spent $1,700,000 to put a hole in the ground: "a very good example—a perfect example—of state ownership of the means of production." But Alberta never did get its royal commission and this tar sands tempest, a dress rehearsal for the contemporary federal-provincial power struggle over energy resources, ended in 1945 when the Dominion's Abasand plant was totally destroyed in another mysterious fire.

It was probably the Abasand fiasco and the fear of further encroachments by Ottawa which decided the Social Credit government to bring free enterprise and the oil sands together as soon as possible. Dr. Karl Clark had argued in 1945 that the province itself should develop the sands; but the Manning administration was having no more of that. It decided to build and operate a pilot plant at Bitumount near the old Fitzsimmons operation, but purely to demonstrate to private investors that commercial development was feasible. It operated for one season in 1949, then was closed down. Clark concluded that all the technical problems in oil sands extraction had been solved. Next the province appointed Clark's longtime assistant, Sidney M. Blair, a founder of Canadian Bechtel (then interested in building pipelines from the tar sands), to look into the economics of production. Blair's 1950 report outlined a fully integrated 20,000 barrel per day operation, from strip mining to market, and estimated that a barrel of upgraded synthetic fuel could be produced for $2.08, transported to Superior, Wisconsin, for another $1.02, and sold for $3.50—realizing a profit of $0.40 a barrel and a five and one-half percent return on investment (see Appendix). The Blair report was given a great deal of publicity, and the next year there followed the Oil Sands Conference and the leasing of the tar sands.

Dr. Clark had posed the decisive question: did tar sands development belong to his own generation or not? Thanks to the long years of work by himself and the Research Council, Sidney Ells and other government scientists, it now seemed as though the issue had been resolved. Development was feasible from every point of view, and the government was doing everything it could to persuade the oil industry that a good profit could be turned in the tar sands. Yet, another fifteen years would pass before the first commercial production of the sands—an event Karl Clark did not live to see. What Clark and the Social Credit government had failed to reckon with was a quite decisive point: it was not in the industry's interest to develop the oil sands before then. An angry Bob Fitzsimmons had predicted that the oil companies would gladly take over the leases and all the technical information available, but only to keep the resource out of production. By taking up the exclusive and cheap concessionary rights granted in the twenty-one-year leases, the companies could keep potential competitors out of the tar sands and simply wait until their own worldwide interests tipped the scales in favour of development. Had the work of the provincial and Dominion governments been continued, development of the oil sands could easily have begun in the 1950s. Under the auspices of the oil industry and a continental oil policy, however, their development was set back nearly two decades.

By the early 1960s most of the prime acreage in the tar sands was under the control of the major oil companies. Once again it seemed as though several extraction plants would soon be constructed. Several smaller companies, among them Sun, Richfield and Cities Service, were interested in improving their crude supply position, and even one or two of the majors were planning developments. The Syncrude consortium took shape about this time around the old Cities Service-Athabasca project planned by Cities Service, Richfield Oil and Imperial Oil (who had banded together in 1958 to buy up some two million acres of permits and leases in the oil sands), together with Royalite Oil, later to be absorbed into Gulf. Shell Oil was also planning a large *in situ* operation. But the government of Alberta was by now fearful that an excess of production from the tar sands would upset the marketing arrangements of the conventional oil industry; thus in 1962 it declared that output from the sands must not exceed five percent of total Alberta production in markets within reach of the conventional oil industry. The Shell and Cities Service applications were deferred in favour of a much less ambitious oil sands project.

Ironically, the company destined finally to begin exploitation of the tar sands was the direct descendant of Bob Fitzsimmons' hapless International Bitumen Company. Now called Great Canadian Oil Sands, the company formed by Lloyd Champion had a lengthy history of involvement with both the Social Credit government of Alberta and the Sun Oil Company of Philadelphia. Champion had escorted three Sun representatives around the Bitumount site back in the 1940s and Karl Clark, who happened to be there as well, found the Sun men "extremely tight-lipped about their trip. When asked whether the oil industry was so concerned about reserves as to be seriously interested in the tar sands, their reply was, 'If Sun can get oil from the tar sands as cheaply as from oil wells, why should we not be interested?' " In 1954 Sun acquired its oil sands leases, including a seventy-five percent interest in the 4,000 acre lease No. 86, holding the rich Tar Island deposit uncovered by Dominion surveyors after the war. In 1958 GCOS contracted with Sun Oil for the right to develop lease 86, while Sun in turn agreed to purchase seventy-five percent of GCOS production (Shell buys the remaining twenty-five percent). In 1962 GCOS obtained permission from the province to build a 31,500 barrel per day (later 45,000 b/d, now 65,000 b/d) integrated mining, extraction and refining complex; then, in a bid to raise financing, GCOS gave Canadian Pacific Railway a fifty-one percent interest in the company. CPR in turn sold two-thirds of its option to Sun and Shell, but Shell and CPR dropped out and as a result Sun Oil acquired what Bob Fitzsimmons began in the tar sands back in 1922.

On September 25, 1967, some five hundred dignitaries gathered north of Fort McMurray on the west bank of the Athabasca for dedication ceremonies at "the world's first oil mine," Great Canadian Oil Sands. All speakers agreed that it was an occasion pregnant with meaning for future generations of North Americans. *Oilweek* later described the scene. "Throughout the endless speeches, a lonely looking old man sat silent and impassively at the head table, huddled deep into a blue overcoat, the collar turned up at the back and rimless spectacles riding down an ample nose." Rumour had it that the old man had taken the decision by himself, overriding the objections of his board and threatening to finance tar sands development with his personal fortune. He was John Howard Pew, eighty-five-year-old patriarch of one of America's wealthiest families, supporter of ultra-right causes and candidates, chairman of Sun Oil Company of Philadelphia, at the time the twelfth largest oil company in the United States. They were a perfect match—Pew and Ernest Manning: two patricians of corporate and political power, both deeply conservative, religious and paternalistic, launching the Athabasca tar sands into the global struggle against Godless Communism on the side of right, free enterprise and the North American Way of Life. "No nation can long be secure in this atomic age unless it be amply supplied with petroleum," Pew advised the assembled crowd. "It is the considered opinion of our group that if the North American continent is to produce the oil to meet its requirements in the years ahead, oil from the Athabasca area must of necessity play an important role." North Americans would never again have to worry about an oil shortage, predicted *Oilweek,* a trifle prematurely. J. Howard Pew had brought into being a petroleum insurance policy for the continent.

It was hardly surprising that the tar sands attracted little interest in a period of over-production of conventional oil. Throughout most of the 1960s the central dilemma of Alberta's oil policy lay in finding markets to absorb the province's surplus production. By mid decade new discoveries had unexpectedly boosted the conventional crude oil life index to thirty-one years, and a large percentage of discovered reserves was being shut-in, in effect restricting production. The 1961 National Oil Policy, devised after bitter opposition by the international oil companies and the U.S. government to the idea of opening the Montreal market to domestic Canadian production, forced domestic producers to find outlets for their surplus capacity in the U.S. export market. The National Oil Policy closed the door on any major development of the oil sands in the 1960s. Alberta's restrictive oil sands policy, enunciated a year after the National Oil Policy and subsequently modified, in effect directed potential developers to find assured export markets beyond the reach of the conventional oil industry before submitting applications to the Oil and Gas Conservation Board (now the Energy Resources Conservation Board). GCOS was allowed to go ahead, but Syncrude and other ventures had to wait until the continental energy equation showed conclusively that additional oil sands production could be absorbed by the U.S. market without further shutting-in of conventional production. Under this arrangement U.S. supply and demand governed Alberta's tar sands policy; hence the 1968 discovery of oil on Alaska's north slope further delayed additional development of the Athabasca deposit by several years. In retrospect, it seems apparent that earlier exploitation of the sands could only have occurred in the framework of an independent national energy policy. Under the policies actually followed they were held in reserve for the day when the United States would need them. In essence Syncrude was originally designed and approved for an export market.

As for Great Canadian Oil Sands' pioneering plant, it has had a troubled and somewhat controversial career since coming onstream in 1967. The project encountered many unexpected breaking-in problems, its operating costs soared above projections and it quickly began to show heavy losses. Before turning its first profit in the second quarter of 1974 GCOS had run up total losses of some $90 million. Naturally enough, these losses have provided excellent ammunition for other tar sands applicants seeking easier terms for their plants. But critics have suggested that these losses and GCOS's high costs are a consequence of learning costs facing any pioneering operation, its lack of economies of scale (its daily production is less than half of the Syncrude generation plants), the company's conservative accounting procedures, and its failure to obtain a price proportionate to the quality of its product. Allegations have been made that GCOS has been turned into a captive producer by its parent, Sun Oil, and that its losses have been artificially created to win royalty remissions from the province and tax concessions from Ottawa. GCOS strongly denies these charges. The company's president, K. F. Heddon, argues that without Sun Oil, ''GCOS today would be a mere footnote in some book. . . . Rather than milk GCOS for profits, Sun Oil in various ways . . . has enabled GCOS to keep operating.'' Neither is it true, says Heddon, ''that Sun Oil purchases crude from GCOS at reduced prices, thus transferring profits to the parent company.'' Sun purchases crude ''in accordance with its sale of product agreement, an agreement based on principles established as far back as 1958, long before Sun had an equity interest in GCOS.'' Sun Oil pays ''the highest price that GCOS can command from refiners on the open market and what we can command depends very substantially on the going rate for conventional crude.'' The GCOS chief argues that the attacks on his company have a sinister purpose: ''Attacks of this nature are most disturbing. Their sole aim is the spread of disinformation which seems designed to do nothing more [sic] than undermine the free enterprise system.''

In 1974 Alberta's provincial auditor looked into these charges and countercharges and gave GCOS a clean bill of health. The audit concluded that first, the selling price of synthetic oil produced by GCOS appears to be established by the ''market place,'' except for that sold under long-term contract to Sun and Shell Oil. Second, GCOS's accounting policies ''appear to have been most conservative,'' though the auditor considered such conservatism justified by the pioneering nature of the enterprise and the uncertainty as to its future.

Nevertheless, the auditor's report did not resolve the crucial issue: what is the true value of the synthetic oil? Since 1971 GCOS has received a small premium for quality, and most of its oil sells for slightly more than average crudes from Alberta's conventional fields. Back in 1950, however, the Blair report determined that tar sands synthetic oil (''desulphurized distillate'') should be worth a price to refiners one-third higher than that for Alberta Redwater crude. Obviously, the government cannot judge such matters on the basis of an outdated report, but it would be interesting to know what kinds of studies have been done on this question since 1950. Furthermore, it is undeniable that GCOS has been a burden to the taxpayer—though it does not approach Syncrude's scale. Between 1967 and 1974 GCOS was remitted $7.1 million in royalties by Alberta and another $6 million in sales taxes by Ottawa. The company's net royalty to Alberta in that same period was a slim twenty-five cents per barrel of sales—which is not exactly a great return to the owners of the resource.

In 1971 the Syncrude consortium finally received approval from the Conservation Board for a 125,000 barrel per day complex on their Mildred Lake lease No. 17, next door to GCOS. The subsequent history of the Syncrude negotiations will be taken up in later chapters of this study. Three other similar mining projects, now estimated at above $2 billion each, are pending or under negotiation as of the spring of 1975. In the lineup behind Syncrude is Shell Canada Limited's proposed extraction operation on lease No. 13, located fifty miles north of Fort McMurray on the opposite side of the Athabasca River. In October 1974 Shell Canada's sister company and fifty-fifty partner, Shell Explorer of Houston, withdrew from the project. Shell's terms are presently under negotiation and there is every indication that the company's demands are even more excessive than those of Syncrude.

Shell's plant could one day be flanked by two other extraction projects. Petrofina Canada Limited, a seventy-two percent owned affiliate of Petrofina S.A. of Belgium is the operator and 35.33 percent owner in the proposed Athabasca Oil Sand Project (AOP), a 122,500 barrel a day complex designated on the Daphne Island block of leases directly north of Shell's site. Partners with Fina in the AOP venture are three American-owned firms: Pacific Petroleums Limited (forty-eight percent owned by Phillips Petroleum Company) also has 35.33 percent interest; Hudson's Bay Oil and Gas (fifty-three percent owned by Continental Oil Company) has 18.8 percent of AOP; and Murphy Oil of New York has the remaining 10.5 percent share of the venture, optimistically scheduled for production by 1982.

The fifth surface mining and extraction plant in the queue is a 103,000 barrel per day project being planned by Home Oil (87.5 percent interest) and Alminex Limited (12.5 percent interest). Home is controlled by Consumer's Gas Company of Toronto and is the largest Canadian-owned oil company. Alminex is a Toronto-based resource outfit controlled by Falconbridge Mines, but Falconbridge itself is controlled by Superior Oil of Houston. Home Oil's plant, if it is built, will be located due south of Shell's site. Shell, AOP and Home have had discussions regarding the possible development of a new town, located north of McMurray, to service their operations. Without doubt the town will be largely paid for by the province of Alberta, should the companies decide they need it.

All of these ventures have been approved by the Energy Resources Conservation Board, which is hardly surprising in view of the board's vision of the province's future. If the ERCB's economic forecasts are ever realized, Alberta will one day be dotted with huge petrochemical plants, coal gasification plants, strip mines, tar sands extraction operations, and myriad other blessings of industrialism—a veritable New Jersey of the north crisscrossed with energy corridors and roads to nonrenewable resources. Whether this blueprint for the future will be translated into reality, however, depends to a very great degree upon the decisions of the world's major actors in the politics of energy and the willingness of our governments to meet their price tag for exploiting our resources. The future of the oil sands is inextricably linked to the larger continental patterns of Canadian economic development.

4   Fortress North America

"Because no one is closer to us, we think immediately of Canada when we discuss the scope and importance of the president's announcement of Project Independence."

William J. Porter, U.S. ambassador to Canada
May 9, 1974

One month following the outbreak of the Yom Kippur Middle East war of October 1973, with the United States deep in the grip of Watergate fever compounded by anxiety over the Arab oil boycott, former U.S. President Richard M. Nixon appeared on American television to prescribe strong medicine, his antidote for the energy crisis.

Nixon named it "Project Independence." The challenge facing the United States, he declared, was to regain the strength of self-sufficiency in energy. This was a key to America's predominance among the nations. "Our ability to meet our own energy needs is directly linked to our continued ability to act decisively and independently at home and abroad in the service of peace, not only for America, but for all nations in the world." Calling for "focused leadership" to achieve self-sufficiency by 1980, Nixon likened his challenge to earlier crash programs to develop the atomic bomb and to put a man on the moon. He went on to promise massive public funding for the exploration of America's remaining energy resources—Alaskan oil and gas, offshore oil reserves, nuclear energy and synthetic fuels from coal and oil shale. A few days later Nixon reiterated his challenge, linking it to rumours circulating in Washington that the "blue-eyed Arabs" of Canada were taking advantage of

49

America's energy plight. The United States, Nixon asserted, should be independent of all oil-producing countries, "including our Canadian friends," by 1976. Canadians "can be pretty tough on us sometimes when they are looking down our throats." This did not mean that the U.S. would not continue to desire the oil of the Middle East or the gas of Siberia or that she would cease energy cooperation with Canada or Latin America. "But it does mean that the United States must be independent in this area, and we can be."

What was behind the new U.S. energy strategy, and what did it mean for Canada? Did Nixon's speeches mark an end to the era of continental pressures on Canadian resources? Had the United States decided to turn its back on Canada's present and future energy reserves? Did Project Independence constitute a turning point in Canadian-American relations?

In the first place, Project Independence represents nothing new or revolutionary in American history. Rather, it marks a partial return to traditional "Fortress America" policies, albeit on a grand scale. The United States has a long history of energy "crises" and most of them have had similar origins and outcomes. Intermittently—and usually when the petroleum industry fears new competition or a decline in world prices—the public alarm has been sounded about a growing, ominous dependence on imports, declining U.S. production and the need for the vigorous development of domestic energy reserves. Traditionally, these warnings have dwelled on the dangers inherent in such a situation for American national security. Energy scares of this kind occurred in the early 1920s and again in the aftermath of World War Two. The response to these panics has typically been to reserve and protect the American market for domestic producers while encouraging the major international companies to acquire foreign markets for their huge reserves of offshore oil. The majors, who hold dominant positions in both the U.S. and international oil markets, have been able to win truly extraordinary measures of support from the American government by linking their own fortunes to the country's paranoia about national security. Consequently, the oil industry has emerged from successive energy panics with its profit margins improved and new legislation on the books protecting its domestic and foreign interests. In the 1950s, for instance, at a moment when the majors were facing severe competition abroad from a group of smaller, aggressive companies, the Eisenhower administration caved in to pressure from the powerful domestic oil lobby and created oil import quotas. The quotas restricted the entry of smaller companies into the U.S. market, kept domestic oil prices up and won the backing of top military and diplomatic policy makers on the grounds that energy protectionism enhanced national security. The same grounds accounted for Canada's special, preferred role in U.S. energy planning. As preferred areas of supply only Canada and Mexico, later Canada alone, were exempted from the import quota system. "It has always been the policy in government," one U.S. official explained in the mid 1950s, "to consider those countries and others in this hemisphere as within the U.S. orbit when dealing with defence questions." That assumption persists to this day.

If America's current energy crisis is not so new, neither—appearances notwithstanding—was Project Independence simply a public relations gimmick, a crass attempt to divert public attention from the Watergate scandals with patriotic appeals for sacrifice and belt tightening. Nor was it a snap reaction to the oil embargo announced weeks earlier by the Arab producers. Undoubtedly, these affected the timing and the overly dramatic delivery of the speeches, but the real origins of this energy policy predate Watergate and the oil boycott by several years.

At the end of the 1960s it was clear to some Americans that their country was facing a serious energy squeeze. Annual consumption of energy had been climbing at an astronomical rate while domestic production had begun to fall off. In 1970 the Cabinet Task Force on Oil Import Control, headed by Labor Secretary George Schultz, noted the increased risk of American dependence on foreign oil supplies and called for replacement of the oil import system with a program of tariff preferences and restrictions. The Schultz report gave special attention to Canada's political stability, concluding that "there may be reason to consider Canada more reliable than any other foreign supplier." The United States should negotiate common or harmonized energy policies with Canada, the so-called continental energy policy, and Canadian oil exports could then be given preferential treatment. Or, as George Schultz later spelled it out, "we know there are large sources of energy in Canada. We would like to be recipients of some of that."

The continental energy deal remains a major theme of U.S. policy; indeed, the need for guaranteed access to Canadian energy supplies has become even more pressing since the Schultz report. More recent projections have argued that the task force badly underestimated the extent of America's energy thirst and the dimensions of the security threat by forecasting that by 1980 the U.S. would be consuming about 18 million barrels of oil per day (b/d), two-thirds of which could be produced domestically. Studies released in 1972 by the Department of the Interior, the Office of Emergency Preparedness and the National Petroleum Council predicted that U.S. oil consumption could reach 25 million b/d by 1980, more than half of which would have to come from imports. Projections to 1985 and beyond show an increasing dependence on supplies of foreign oil, and only the surplus productive capacity of the oil rich powers of the Persian Gulf, it was clear, could make up the growing difference between demand and supply over the next ten to fifteen years. Because these estimates originate with an industry notorious for its misuse of statistics and which may be attempting to stampede government into concessions it badly desires, they should be treated with great circumspection. However, although the true extent of the dependence may be debatable, it is beyond dispute that America will be unable to satisfy her own energy appetites and that this is genuinely regarded in Washington as an intolerable situation.

"The United States simply cannot afford an ever increasing overdependence for its oil supplies on a handful of foreign, largely unstable, countries. Otherwise, its security—and that of its allies—as well as its prosperity and its freedom of action in foreign policy formulation will be in jeopardy." This is the lesson drawn from America's supply-demand crisis by one of her foremost authorities on energy questions, prominent New York oil consultant Walter J. Levy. Levy, whose special interests in Canada's oil sands will be discussed later in these pages, has had a major influence on charting the course of postwar American energy policy. His approach to past and present energy problems has been so representative of official thinking that his views deserve far more public attention.

Walter Levy was described by the New York *Times* in 1969 as "unique," "the dean of oil consultants," and "the confidant of presidents, shahs and sultans." The newspaper reported that "there are few, if any, major oil controversies in which Mr. Levy has not acted as a consultant." He has worked "as an advisor to most of the major oil companies, most of the important consuming countries and many of the large producing countries. He has worked on all sides of the powerful and sensitive oil industry without losing a reputation for objectivity. Countries have asked for Mr. Levy's opinion even though they were completely aware that he was a consultant to the companies with which the particular nation was negotiating." In 1968 Levy, who worked with the Office of Strategic Services in World War Two and later served on special missions to Iran and the Middle East, was awarded a special plaque by U.S. Secretary of

State Dean Rusk: "In grateful appreciation for your invaluable contribution to the welfare of the United States." The New York *Times* remarked in 1974 that Levy "has the ear of Dr. Henry Kissinger," current Secretary of State, and U.S. energy policy certainly bears the stamp of Levy's ideas. At one time or another, and often simultaneously, Levy's consulting firm has worked for the leading oil companies, the European Economic Community, the World Bank, the United Nations, and the governments of a wide group of oil-producing or consuming countries, including Canada. His earliest contribution to Canadian energy policy came in the late 1950s when his company consulted for the Borden Royal Commission—"thinking through the problems of an oil policy for North America," *Business Week* reported. One knowledgeable source flatly states that Levy fathered Canada's National Oil Policy of 1961, that peculiar continentalist compromise which left eastern Canada consuming Venezuela crude while western Canadian producers sought American buyers for their surplus oil. Here indeed was an "oil policy for North America." Levy has never deviated from his position that a continental approach to energy matters is sound strategy for both the United States and Canada, a fact which makes his current $60,000 annual consulting fee from the government of Alberta all the more intriguing.

Levy's immensely influential reputation with governments and companies—one source calls him "the Kissinger of oil"—is said to derive from his ability to define negotiable issues and to find the gray area in which conflicts can be resolved. He himself believes that the key to his success is his impartiality: "My whole business has been built on a reputation for objectivity. If I waver from this I have nothing." There is no reason to question Levy's integrity or his sincerity on this point; but it must be noted that such "impartiality" has very definite limits. Levy shares the philosophical frame of reference of his oil company clients and one of his firmest beliefs is that energy development is best left to the experts of the international petroleum industry. *Time* magazine recalled how in 1959 Levy attempted to dissuade India's Prime Minister Nehru from being too ambitious in exploring for petroleum with Indian resources. Said Levy, "for every oil well you drill, 1,000 Indians will have to go without an education. Your resources are inadequate to do everything you want. So let foreign interests do the drilling." "Levy's advice helped to temper Indian policy," remarks *Time*. Levy made the same crucial argument in a World Bank study circulated among underdeveloped countries in 1961. Michael Tanzer, author of *The Political Economy of International Oil and the Underdeveloped Countries,* comments that "the reader of this report comes away with the impression that the worst approach to "the search for oil" is for the government to undertake oil exploration on its own." Levy's World Bank report, with its heavy emphasis on the tremendous "risks" and "expenses" involved in petroleum exploration and development, starts from an *a priori* presumption that oil is the business of business and goes on to document a case against governmental involvement. Left out, naturally enough, is any discussion of the case for such involvement, of the social, economic and political advantages of the public control and ownership of vital energy resources. As we shall see, Levy has been giving precisely this kind of "impartial" advice to the Lougheed government in his recommendations for developing the oil sands.

Levy's analysis of the current crisis in international energy and his proposed solutions provide some important clues for understanding the various strands of American policy. His anxiety about the growing gap in U.S. supply and demand projections and the shift in the power centre of world energy politics toward the Arab countries stems from a series of related concerns. In the first place, over the next decade or two only the oil-producing nations of the Middle East have the potential surplus production to help the United States make up its oil deficiencies. Second, this "ever increasing over-dependence" on "unstable" Arab nations poses a serious threat to American security and U.S. foreign policy. Third, the massive cost of oil imports imperils the balance of payments position of every oil-consuming nation, including the United States. Fourth, the accumulation of tremendous financial resources in a few Arab hands could gravely disrupt the already shaky international monetary system or result in Arab takeovers of important American industries. Fifth, the international oil companies have lost much of their power to bargain effectively for the interests of the oil-burning, industrialized world and are progressively being forced into partnership roles with the oil-producing countries. All of this adds up, in Levy's highly political analysis, to a fundamental threat to the interests of the United States and the rest of the "Free World." His solution? "To mitigate the dependence of the U.S. and to confine it to manageable proportions for the short-term as well as for the longer range, a coordinated approach to energy policy is absolutely essential. This would have to take account of all relevant factors conducive to improving the North American domestic supply situation from the traditional sources— oil, gas, and coal—and from new sources for liquid and gaseous energy such as oil from shale and tar sands, synthetic oil from coal liquefaction, natural gas from coal or oil, and electric power from nuclear sources. For the U.S., as a major world power, it is essential to take all such necessary steps because our political, strategic, and economic security depends on the availability of fuel from safe and dependable sources." Through Levy's eyes, "North American domestic supply" is a key to the security of the U.S. as a world power.

The rapid exploitation of North America's remaining energy resources, encouraged by much higher prices and government funding, would be augmented by two other essential policies. First: "We should develop a North American oil policy which would, in fact, coordinate our energy position with that of Canada." Second, and on a far grander scale, the U.S. should try to bolster the faltering positions of the big oil companies and defend the interests of the oil-consuming, industrialized nations by backing what Levy calls "a coordinated Atlantic-Japanese energy policy." The collective muscle of the developed, capitalist countries should be thrown into the power struggle for energy in an attempt to create a "countervailing power" to that of the Organization of Petroleum Exporting Countries. The consuming nations should avoid bilateral, "go it alone" deals with OPEC because such practices encourage government to government oil transactions and undercut the traditional intermediaries, the oil companies. An Atlantic-Japanese policy would encourage and guarantee investment in new unconventional energy sources (such as the oil sands), develop programs for optimum diversification of supplies, set up emergency contingency plans for dealing with embargoes, deal with capital movements and balance of payments problems, and review the powers available to the large, capitalist countries for dealing with OPEC. The ultimate objectives of Levy's policy are: to improve the world power position of the U.S. and strengthen the hand of the big oil companies; to erode OPEC's unity; and to integrate the western and Arab economies to such a tight degree that oil embargoes would become self-defeating. America's top energy consultant is not called "the Kissinger of oil" for nothing.

This strategy is not unique to Walter Levy, though he has probably been its most articulate proponent. Rather, it emerges from a consensus in the general thinking of the industry and the U.S. government on recent developments in international oil. Perhaps this is why Levy's strategies for reasserting American power—the accelerated development of high cost, secure energy sources; continental energy arrangements with Canada; and energy cooperation, or multi-lateralism, among the developed, capitalist countries—so closely parallel the actual policies of the U.S. government and the leading oil companies. The whole package is a political design for improving U.S. political and corporate power, and this includes Canada's designated role as an energy-abundant, docile ally willing to supply fuel and diplomatic backing for America's struggle to recapture her slipping hegemony.

It is within the context of this strategy that Walter Levy's advice on developing the oil sands and America's interest in Canada's future energy reserves must be understood. Seen from this global perspective, the interests of the oil majors and the national security interests of the United States coincide around the need to diversify out of Middle East oil into new, high cost, politically secure energy forms. As part of the strategy of diversification, Canada's future fuels—the oil sands, the Arctic and offshore oil and gas, the uranium—should be developed and marketed with those of the U.S. in a joint "Fortress North America" energy pool. But Canada is also being asked to provide support for the policy of multilateralism, the attempt to create a U.S.-led bloc of developed capitalist countries as an alternative to bilateralism and as a countervailing power to OPEC. Canadian diplomatic influence, in other words, is to be used to help shore up the slipping world position of the United States and to provide support for the great international oil companies.

Project Independence then is not simply a major program to improve America's energy supply projections. Rather, it is best understood as one move in a much broader power play whose objective is to strengthen the world economic and political interests of the United States relative to the OPEC bloc and her major capitalist rivals, Japan and Western Europe. Whether the U.S. tries for a very high degree of energy self-sufficiency may therefore depend on the kind of success she has in dealing with these rival power centres. If multi-lateralism fails and the French style of dealing directly with the oil producers catches on, and if the OPEC countries continue their program of increased taxation and phased nationalization of the oil industry, then the U.S. could conceivably abandon attempts at energy cooperation among the consuming powers, make some oil supply deals of her own with Saudi Arabia and Iran and launch a truly serious commitment to total self-sufficiency in the next fifteen to twenty years. Yet, there are immense risks and difficulties in such an approach, and these make it unlikely that ex-President Nixon's zealous version of Project Independence—to make the United States "totally independent of any outside source of energy"—amounts to much more than an implied threat to countries like Canada to fall in step behind the American drummer.

On paper Project Independence certainly looks impressive. Theoretically, all the U.S. has to do is boost its total energy production from the actual 1973 equivalent of 30.5 million b/d to an equivalent of 42.5 million b/d by 1980, while cutting back on the growth in demand by another five million b/d oil equivalent. This is to be accomplished through a concerted effort to bring on stream a variety of unexploited, potential energy reserves. The blueprint includes a major leasing program for mineral rights on federal lands and offshore; the deregulation of natural gas prices; completion of the trans-Alaskan oil pipeline, building of a gas pipeline down the Mackenzie Valley to carry Alaskan gas; stepped up exploration and development of offshore oil reserves; leasing and development of oil shale lands in Colorado, Wyoming and Utah; heavy increases in coal production and the attempt to extract synthetic fuels from coal gasification and liquefaction; construction of new refineries and superports; speedy approval and construction of new nuclear power plants and development of new generation fast breeder reactors; and research into twenty-first century fuels such as nuclear fusion and solar energy. The U.S. government has committed billions of public dollars to subsidize the effort; and because all of the actual development will be done by private industry, it must be understood that Project Independence involves a major transfer of public funds to the coffers of the energy companies. The government has also promised to minimize environmental restrictions, a crucial guarantee in view of the massive ecological costs of Project Independence. Implied too in the plan are, first, a commitment to permanently higher energy prices and profits ("the age of cheap energy is over") and, second, a hidden guarantee to underwrite the entire cost of the program, should it prove a commercial non-starter. As James Laxer rightly puts it, "the new American energy strategy will result in a massive redistribution of income from society at large to the oil companies." Project Independence, whose ultimate cost has been estimated at $510 billion, could be a corporate bonanza, a rip-off unequalled in history.

But, as a sceptical *Wall Street Journal* has noted, "patriotism and profit lures alone can't bring forth fuel." Even without the administrative paralysis and political disunity created by Watergate, energy self-sufficiency is not something which can simply be willed into existence. Speeches, blueprints and theoretical reserves are no substitute for the gas in the tank—a point Canadians also have yet to learn. The *Wall Street Journal* has been keeping a watchful eye on the progress of Project Independence, and the reports have been utterly pessimistic: "In down-to-earth reality, a series of bottlenecks unprecedented for peacetime is hindering plans for tapping additional sources of energy. The obstructions range from a labyrinthine array of restrictive government regulations to shortages of such necessities as oil drilling equipment, coal hopper cars and water for fuel-processing plants." One coal company executive threatens that Project Independence is "dead, buried and mortified unless someone at the top in Washington does something," but it is doubtful that anyone could do enough to bring the plan to realization by 1985, let alone 1976 or 1980. The bottlenecks, vital shortages of equipment and labour, and the long lead times required to bring on stream new energy resources and to bring new technologies into commercial production make it certain that the U.S. will remain substantially dependent on foreign energy supplies for years to come. For instance, the vast oil shale deposits are most unlikely to make any significant contribution to U.S. energy supplies before the 1990s. Synthetic gas and oil from coal conversion are also unlikely, in the assessment of the *Wall Street Journal,* to "make more than a piddling contribution to the nation's energy pool for a decade or so to come." Technological obstacles, shortages of large amounts of water needed to extract the fuels from shale and coal, environmental opposition (the emergence of a "not in my backyard" brand of politics), and shortages of engineers and skilled labour—all these make the synthetic fuels industry an unlikely instant solution to the U.S. energy crunch. Similar kinds of problems have been plaguing the nuclear power industry—licensing delays, equipment failures, labour shortages—and most plants due to be operating by 1980 are at least two years behind schedule. The moral appears to be that Project Independence is easier said than done.

Beyond these very substantial barriers to any easy transition to energy self-sufficiency, the move also involves formidable economic and political risks—risks which some influential Americans already regard as unacceptably high. How, for instance, will the United States be able to protect its high cost, synthetic fuels industries from competition with cheap foreign oil without the burden of other major costs? The entire viability of the U.S. energy substitution program— and Canada's oil sands as well—is dependent on the world price of oil staying above a certain floor, or minimum level (the higher the price of oil, the more attractive energy substitution becomes), but it cannot be guaranteed that this will happen. Some economists now confidently assume that oil prices will stay high, but they are completely ignoring the fact that world oil prices are artificially determined by considerations which are as often political as they are economic in nature. The U.S. Senate Committee on Finance, in a 1973 report titled "Fiscal Policy and the Energy Crisis," poses this highly suggestive question: "At present the Arab nations can charge $9 to $14 a barrel because our existing productive capacities are insufficient to supply our own needs. But if we bring on new production, which may involve costs of $5 to $7 a barrel, and the Arab nations then drop the price to $4 a barrel, where will the American producer stand?" This question is particularly relevant in light of some evidence that OPEC's largest oil producer, Saudi Arabia, wishes eventually to push posted prices below their current level to ensure that alternative forms of energy will not be so commercially attractive. Within OPEC Saudi Arabia is today certainly in a minority, but no one can guarantee that her view will not prevail, particularly if the drive for U.S. self-sufficiency begins to gain momentum, or that OPEC might not disintegrate because of such conflict or the impact of current growing oil surpluses in the world. The American energy industries are well aware of this risk and they and their political supporters fully intend that the risk will have to be underwritten by the U.S. taxpayer. Senator Henry Jackson, one of Project Independence's strongest backers, argues that the oil producers will inevitably try to forestall the move to self-sufficiency by cutting their prices. "For this reason, important high cost sources of domestic oil and gas, and facilites for making liquids or gases from coal or oil shale, may not be developed without some form of governmental intervention in the market to guarantee domestic prices, in the form either of import restrictions or direct or indirect subsidies." One possibility is an oil tariff wall which would keep cheap foreign oil out of the U.S. market, but this could leave American manufacturers paying for higher energy costs than their European and Japanese competitors. Another suggestion, a guaranteed price floor for synthetic fuels, would have U.S. taxpayers underwriting the risks of Project Independence to the tune of $100 billion! Such could be the price of national security, warns the *Wall Street Journal,* but others wonder if the price is worth paying.

This problem is one our oil sands enthusiasts must also stop to ponder. If the oil sands are developed, as industry and government presently intend, for the "world market" price in competition with U.S. synthetic fuels, what happens if at some future date the Saudi Arabians and other oil producers attempt to undercut Project Independence? A major penalty we may ultimately pay for developing our future energy resources privately and according to continental models is that they will then be vulnerable to the kind of Arab squeeze Senator Jackson is predicting—sideline hostages in a showdown among the energy powers. And, of course, the oil sands industry will use this hypothetical threat to demand that Canadian society underwrite its future viability. If the oil sands were well into commercial production when the threat began to materialize, then Canada might be confronted by a choice between massive subsidies to foreign oil companies or an oil sands collapse which would turn Fort McMurray into a ghost town.

Even more troubling to some prominent Americans is the worry that an excessively nationalistic approach to energy might cut across the goal of multilateralism and trigger similar go it alone policies in Western Europe and Japan. This might stimulate the kind of government to government oil deals that companies fear, strengthen OPEC's bargaining hand, and invite protectionist policies which could hurt other American investors. Leading U.S. manufacturers and politicians like Senators Edward Kennedy and Edmund Muskie have described Project Independence as "isolationist" and a Ford Foundation study terms self-sufficiency a "simplistic overreaction." Exxon Corporation, the world's biggest energy company, is clearly concerned about the fate of its overseas investments as well as its future role in U.S. energy. Former Exxon chairman J. K. Jamieson argues: "Even if we agree on the need to achieve the greatest possible energy self-sufficiency for the United States, we must be careful how we use the term. I believe total self-sufficiency, in its rigid and most narrow sense, is probably beyond reach. Further, it is my view that neither the United States nor any other consuming nation should commit itself now to such an excessively nationalistic goal." Calling for "a high degree of international cooperation" in energy matters, Jamieson wants the United States "to lead in solving the energy problem." America should "move forcefully toward decreased dependence. Such a policy is not a prescription for a retreat into narrow isolationism, but rather a means by which the United States will be able to cooperate most effectively with other nations in meeting the world's energy needs."

Exxon's corporate strategists argue that Project Independence can and should be viewed as part of "Project Interdependence." Addressing a Vancouver symposium of the Conference Board in Canada in May 1974, James W. Hanson, Exxon's chief economist, spelled out his conception of self-sufficiency. "All governments of the principal oil-consuming countries should minimize barriers to the movement of capital, technology, and energy supplies among themselves. For each to insist upon a high degree of national energy self-sufficiency rather than seek an acceptable degree of composite energy self-sufficiency could lead to investments in highly uneconomic local energy sources at the expense of more economic sources in one of the other countries, e.g., United States coal, Athabasca oil sands, and Arctic oil and gas. The U.S. and Canada are thus in a position to render a signal service to the cause of Free World political and economic cooperation if exports of these resources are not restrained." Such an outlook closely complements Walter Levy's call for a coordinated Atlantic-Japanese energy policy and is in line with Henry Kissinger's efforts to create an energy action group among the principal consuming countries during the 1973-74 winter. Whether it addresses itself to Canada's long-term interests, however, is another question entirely.

With the energy crisis threatening to metamorphose into a full-blown international monetary crisis, the need for multilateralism among oil consumers is felt to be more pressing than ever. The Italian economy has already been seriously damaged; those of France, Britain, possibly Japan, are under severe pressure because of the tremendous pressures on the balance of payments of these nations caused by the sudden four-fold jump in the cost of imported oil. The U.S. economy has also been feeling the shock of high cost oil imports and American policy makers have worried about the fantastic trade surpluses piling up in the OPEC countries. Fears have been expressed about the possibility of the Arabs attempting to buy out the "downstream" U.S. operations of one or more of the major oil companies or of their acquiring controlling assets in such huge and symbolic corporations as General Motors. "It is most unlikely," writes Walter Levy (who does not hesitate to advise others to accept foreign ownerships), "that the United States, or any other developed country, would permit continued massive foreign investments on a scale that could conceivably result in foreign takeovers of important companies and industries." Ingenious schemes have been devised for "recycling" the wealth piling up in Arab hands—"recycling" being the euphemism for "getting it back into responsible hands." One favourite proposal has the oil producers turning over a large part of their revenues to the poorest Third World countries who could then buy more goods from the industrialized countries. Not surprisingly, the Arabs and the other oil-exporting countries find this sudden concern for the fate of the Third World lacking in altruism and they have not been lining up to "recycle" their new wealth. The real salvation of western capitalism may lie in the fact that the elites who run countries like Iran and Saudi Arabia rely on the U.S. for assistance in maintaining their domestic and regional positions of power, and it is therefore not in their interests to bring about a collapse of the international monetary system.

American energy officials are well aware that total self-sufficiency within one or two decades is unreachable: they now speak of achieving a "capacity for self-sufficiency" by ending a dangerous reliance on "insecure" (Arab) sources. But they don't envision stopping imports from "friendly" suppliers like Canada or Venezuela: one official aptly described American intentions when he spoke of the need for "hemispheric self-sufficiency." This means that as the U.S. moves away from dependence on Middle East oil, the pressures on "friendly" suppliers to share their depleting conventional and undeveloped energy resources will escalate. The threat to exclude uncooperative countries from the long-term "benefits" of Project Independence will be used for maximum leverage—a return to the tactics the U.S. used in 1970 when she imposed an import quota on Canadian oil in an attempt to pressure Ottawa into a continental energy treaty. In the spring of 1974 an official with the U.S. Federal Energy Office told a gathering of Canadian resource executives that Project Independence did not indicate a total withdrawal from the world petroleum market or a refusal to import energy. But for the U.S. to adopt policies which include the use of imported fuels, the knowledge of future volumes on which it could rely were critical. Without this long-term planning capability, investment decisions might be made in other directions. Ever sensitive to Washington's moods, *Oilweek* drew the right conclusion: "It was a velvet glove, rather than a mailed fist approach, but the implications were clear. If the U.S.A. was not assured of stability of supply by Canada, it would not be interested in seeing U.S.A. monies used to develop our resources."

Far from implying a sudden turn away from a continental approach to energy, Project Independence actually promises an intensification of pressures on Canadian resources. The real message that Richard Nixon was conveying to Canadian government officials and businessmen in November 1973 was this: We are going ahead with the development of tomorrow's energy resources, with or without your cooperation. If you want our assistance in developing your own future resources—our markets, investments, materials and know-how—then give us in return a commitment of secure, uninterrupted energy supplies.

It was the same old message. And the message got through.

5   Thinking the Unthinkable

"From the standpoint of both U.S. and Canadian policies discussed above, there is an advantage to moving early and rapidly to develop tar sands production. For the United States, early development of the tar sands could contribute to the availability of secure North American oil supplies over the critical period before its own long-run efforts to develop conventional and synthetic oil might begin to pay off. For Canada, the establishment of early and substantial volumes of tar sands production could be essential to maintenance of Canada's export potential, providing an offset to Canada's rising volumes of oil imports."

"Emerging North American Oil Balances:
Considerations Relevant to a Tar Sands Development Policy"
Walter J. Levy
Consultants Corporation
February 1973

Fueled by the restless, profit-seeking urges of modern industrial capitalism and the security obsessions of a troubled, overextended empire, an unprecedented assault against Nature's remaining raw wealth is being carried out by governments and corporations in the United States and other energy hungry countries. "I would colonize the moon if I could," brayed Cecil Rhodes, but not even that archpriest of expansionism dreamed of forced transformations and upheavals on the scale envisaged by the architects of Project Independence.

    The drives for national security and profit, greed layered over fear, propel what American journalist James Ridgeway aptly calls *The Last Play.* The last play is the struggle by the great oligopolists, the world's energy companies, to corner the market on the world's coming energy needs, buying up and controlling today what will be in demand tomorrow. It is a process which is already far advanced and which literally knows no geographical and few political limits. The quest for energy is a great homogenizer. It cuts through national and ideological shells like a knife penetrates soft butter, demolishing old-fashioned concepts such as sovereignty which might impede "development" and "progress." So long as the standard discounted cash flow analyses project a good rate of return on investment, the energy juggernauts will go anywhere and deal with any kind of political system in their search for the globe's future fuels. Across the world, dime-a-dozen cardboard companies like Syncrude and Canadian Arctic Gas spring up in response to specific needs, jointly fashioned by members of the energy cartel in the interests of "constructive competition," later to be discarded like rusted oil drums when the resource that gave them life is depleted. They come with power and promises, offering economic growth and development in return for the right to drill a hole here, lay a pipeline there, but in the end, when the holes are dry and the pipelines empty, they always go away. And, as James Ridgeway and others have shown, the thrust of the last play increasingly points in the direction of the stable, traditional bastions of white colonialism, particularly Canada.

Like every wellheeled American extravaganza, Project Independence has its share of branchplant Canadian enthusiasts. Although it is clear that most Canadians, particularly the underprivileged and dispossessed, will still be watching from the sidelines when the energy companies pack up and will have reaped little or no benefit from their country's participation in the frantic, uncontrolled scramble for tomorrow's energy resources, there are others who seem only too willing to have Canada "cash in" on America's "energy crisis." These men, not content to have engineered Canada's wholesale resource sell-outs of the 1950s and 1960s, are superbly placed to implement a world view which is profoundly colonial; indeed, they are the ranking members of Canada's economic and political elite. Far removed from the vulnerable hinterlands which must bear the heavy social and ecological costs of the "development" they preach so passionately to each other on the cocktail and luncheon circuit and through the financial pages, Canada's continentalist elite sense a rare opportunity in America's growing fears about her imperial power. As a preferred source of supply, they argue, Canada now has the unique chance to benefit from U.S. efforts to improve its energy supply and demand equation. Under their game plan—an updated, futurist-oriented variation on the old theme of continental energy sharing—the outlines of Project Independence would be drawn around the North American continent and Canada's energy reserves rapidly exploited in response to U.S. demand. Good colonials to a man, their vision is blinkered, their imagination blunted, by their inability to conceive anything but continentalist solutions to continentalist problems.

Consider the views on our future energy resources of one of the more intellectual, articulate members of the Canadian establishment, Ronald S. Ritchie. Ritchie is a former senior vice-president of Imperial Oil Limited, Canada's largest energy company, the first chairman of the "think tank" Institute for Research on Public Policy, a member of the prestigious Club of Rome, an unsuccessful Conservative aspirant in the 1974 federal election campaign, and later assistant to the leader of the opposition. Ritchie's ideas are worth considering in some detail, first, because they appear to be representative of the thinking of many of Canada's leading business executives and political figures, and second, because they lend insight into what can only be called the continentalist fixation.

Speaking in the spring of 1974 to the *Financial Post*'s aptly named symposium, "Our Disappearing Resources," in Toronto, Ritchie argued that contrary to the fashionable wisdom handed down from Malthus to *The Limits to Growth,* the world will never really run out of resources. "We do not ever actually use up the world's energy or its materials. We simply change their forms." Technology adapts continuously to new energy and material resources and there is "no reason to suppose that this process cannot continue indefinitely." We can anticipate replacing one nonrenewable resource with another as technologies and costs change. Substitutes will be developed and the resources available, even fossil fuels, will multiply. What this means is that although Canada's "significant, proved and potential supplies of resources attuned to the world's present consumption patterns and technology will continue to be economically valuable for some decades to come," we should not "neglect the risk that at some time in the future they may not be resources in an economic sense because of changes in the world's technology and demand patterns." This risk may be relevant to "some early policy decisions." Or, as a less sophisticated federal cabinet minister was once heard to wonder, why should we sit on our oil when we'll be heating ourselves with the sun's rays in a few years?

Ritchie spelled out the practical implications of his argument in an assessment of energy issues "from a Canadian perspective" in a recent issue of *International Perspectives*. Arising out of the energy crisis, as he read it, was "the larger need" to direct "North American efforts" to developing major alternative sources of energy. "North America has a special place in the equation of alternate energy sources." The United States has "huge fossil-fuel possibilities" in the form of offshore conventional oil and gas, oil shale and coal. Canada has offshore and Arctic potential, the tar sands and the heavy oil deposits of Alberta, plus significant coal possibilities. "The new situation in world energy markets provides the economic incentive for efforts to develop all of these potential resources, thus giving North America the opportunity to become at least self-sufficient and to cease being a competitor for the energy resources of the eastern hemisphere. From Canada's point of view, it represents the opportunity to convert resources which were only theoretical hitherto into economic assets which can be developed and marketed to the advantage of Canada's national income future." Canada, in other words, can and should cash in on Project Independence.

It would be difficult to find a specimen of thought more representative of all that is wrong with Canadian energy policies than this. Nowhere does Ritchie's glib, facile optimism give pause even to consider the heavy human costs and the ecological degradation we should have to accept as the price of his technocratic solutions. Energy substitution is an extremely arduous, time-consuming and expensive process which has very definite limitations. Should Canadians run short of oil sometime in the early 1980s—as we almost certainly will—we will not awake one miraculous morning to the hum of nuclear power furnaces, our houses lighted by solar energy, with electric powered cars lubricated with oil from the tar sands sitting in our driveways. There can be no magical, swift or painless transition from a fossil fuel economy to whatever succeeds it. When we run short of oil we will quite simply have to get more—at virtually any price—or our economy will grind to a halt and we will freeze to death: yet we have continued to export between forty and fifty percent of our depleting reserves of conventional oil and gas, the best and cheapest and most accessible we will ever own, on the mindless assurance that technology will always provide the answer. Even over the long run, energy substitution and replacement will be immensely difficult, extremely costly, socially and ecologically disruptive; and for each problem resolved by a new technological solution several fresh ones will be created. These facts alone would seem to render absurd any pace of new energy development "attuned to the world's present consumption patterns and technology"; indeed, they argue for precisely the opposite choice—a cautious policy of energy conservation and development aimed at national self-sufficiency. Far from depleting in worth by remaining undeveloped for a time, our resources would increase in value lying in the ground if they eventually help us avoid or reduce some of the tremendous costs of replacement and substitution. As for Ritchie's promise of "North American self-sufficiency"—yet another euphemism for continental energy arrangements—we might recall that Canada is only confronting potential energy shortages in the 1980s because her conventional oil and gas reserves were exploited according to such a design.

What is most intriguing about Ritchie's analysis is what it omits. In our society technologies do not fall from trees or spring up from the soil; they tend to come into the world wearing a corporate trademark. It is a point Mel Watkins makes when he quips, "By the time we get solar energy we'll find I.T.T. owns the sun." Ritchie may be forgiven for assuming that what is good for oil companies is good for Canadians, but the rest of us should ask one simple question: who benefits? Who benefits if Canada does try to cash in on America's energy squeeze and we toss our future resources into the continental market? Who benefits from an accelerated, crash program of development of the oil sands and the reserves of northern Canada? The energy companies have every interest in seeing Canadians hustled and stampeded into long-term commitments of our unused conventional reserves and our undeveloped resources to the "world market." Indeed they argue, like Ritchie, that unless the large U.S. market is available then the proper "economies of scale" cannot be achieved and Canada's resources will remain "theoretical" instead of "economic." Canadians have been swallowing this economies of scale claptrap for much too long, for much too high a price. The fact of the matter is that the Canadian market for energy is growing so fast that even an overly rapid pace of oil sands development will not be able to meet domestic oil demand.

There is, in short, absolutely nothing preventing Canadians from developing our future resources as we see fit, at a pace suited to domestic needs with the benefits accruing to Canadians instead of to multinational corporations. Nothing stands in our way, that is, but the self-serving logic of the oil industry and our own political mismanagement.

Ronald Ritchie's myopic, technocratic perception of the world might be dismissed as merely eccentric were it not for the depressing reality that his views are representative of the thinking of those who hold the key governmental and corporate jobs in Canada. What

Ritchie articulates is the dominant ideology of Canada's power elite. It is an ideology profoundly colonial in nature: a borrowed, alien vision of society and the world environment. Thus it is not surprising that when the Canadian elite peers into the future, it does so through the same distorted "Made in U.S.A." lense its uses to understand past and present events.

Addressing the aforementioned "Disappearing Resources" conference in early 1974, the former federal minister of Energy, Mines and Resources, Donald Macdonald, assured the nation's top executives in the resource sector that "we in Canada are commencing a new round of resource exploration. We are anticipating that it will be concentrated in commodities that have not hitherto been systematically explored, such as uranium, or in parts of the country that have not hitherto been accessible to exploration, for example New Quebec or in the Arctic." Echoing Ritchie's determinedly anti-Malthusian perspective, Macdonald argued that "the real constraint which we are faced with is not a shortage of supply but rather the question of price." As international prices escalate and new technologies are developed, the international supply will increase: the only people who have to worry about "limits to growth," it would seem, are those who cannot afford the price. Macdonald is predicting that "the production of minerals in Canada (excluding fossil fuels) will triple by the year 2000." His former deputy minister Jack Austin, now a member of the Senate, has been saying much the same thing, adding only the reassuring notes that most of this three-fold increase will be for export markets and that in the minerals sector "foreign investment will continue to climb in absolute terms." When a true hewer of wood and drawer of water gazes into his crystal ball, all he sees is forests of wood, oceans of water.

Apply this technocratic continentalism to the policy choices we confront in the oil sands and a standard recipe for externally determined development can be predicted with depressing familiarity. In place of the Alberta civil servants' call for using the sands to reverse the historic trend of foreign control of Canadian resources, we could expect to hear yet another rehash of the old "economies of scale" argument. In lieu of their convincing logic that development be dictated by Canadians for the benefit of Canadians, we could anticipate warnings against a static, narrow view of energy development. Against the cautious, socially and ecologically-minded approach of these Albertans, we could expect contrary advice about the advantages of rapid oil sands development. In place of warnings against allowing the global interests of the multinational oil corporations to set the terms and pace of development, we could expect to hear of the need to utilize the capital, markets, and "know-how" of the big energy companies. Unfortunately, this continental blueprint is far from hypothetical.

When Peter Lougheed's cabinet hired Walter J. Levy Consultants of New York to provide the government of Alberta with information on international energy trends and advice on questions such as oil sands development, they guaranteed themselves a continentalist counterpoint to the nationalism of their civil servants. We have seen how Levy's perspective on world energy politics is strongly conditioned by his concern over OPEC's growing assertiveness and his desires to reassert U.S. Free World power and strengthen the bargaining postures of the international oil companies. His three-part policy—measures to improve North American supply, continental energy arrangements with Canada, and cooperation among Atlantic-Japanese consuming countries—closely resembles the actual policies pursued by U.S. administration in recent years. Another major theme of Levy's thinking, as we have shown, lies in his conviction that new energy development should be left in the hands of industry: government should be discouraged from direct entry into the energy business, restricting its interests to matters such as tax collection or passive participation. Levy shares the philosophical frame of reference of his oil industry clients, and this fact must strongly qualify his reputation for impartiality.

Why, then, does Alberta pay Levy's firm a $60,000 per annum consultant's fee? Bill Dickie, the province's former minister of Mines and Minerals, has defended the hiring of Levy on the grounds that no one in Canada has the equivalent expertise on energy matters. This is simply untrue. It is undeniable that Levy enjoys a unique reputation and has access to a great many influential, first-hand contacts in the world energy field; but with the access comes a perspective, or set of values and attitudes, which does not and cannot reflect the interests and priorities of a minor actor in the international energy scene such as the province of Alberta. Levy looks out on the world through American eyes, his interest is in the global balance of power and in containing what he believes is a fundamental challenge to that balance. Moreover, the "expertise" of his firm's reports, all of which use secondary sources available to anyone, is highly debatable. Only second-hand statistics and estimates on reserves, pricing, supply-demand patterns, etc., are used; no primary data is generated, and there is a heavy reliance on oil industry sources. Levy's forecasting techniques are remarkably out of date, almost primitive, amounting to little more than a regurgitation of other people's projections, again with a strong bias for industry figures. There appears to be little data in Levy's two-volume report on the oil sands which could not have been gathered by an efficient graduate student with access to the same published sources. And far more sophisticated long-range forecasting techniques are available at any number of Canadian universities. Right next door to the government in Edmonton, the University of Alberta contains a large pool of energy expertise which Mr. Dickie overlooked on his way to New York. In light of this, the suspicion arises that the Lougheed government may have been looking more for intellectual reinforcement than expertise when they hired Walter Levy to advise them on oil sands development. If so, they found the right man.

Reading the recommendations of Alberta's civil servants and those of Walter Levy's firm on how to develop the oil sands, one is immediately struck by the contrasts of the two. First a view from the hinterland, then a report from the metropole looking down the imperial tube. As the Levy title indicates—"Emerging North American Oil Balances: Considerations Relevant to a Tar Sands Development Policy"—the underlying assumptions are explicitly continentalist in nature. Most of the discussion focuses on trends in American energy policy and their relevance for the oil sands. Great emphasis is placed on the growing dependence of the United States on foreign oil and on American concern over the security of its imported oil supplies. Written in February 1973, months before the Arab oil embargo, Levy's report stressed how the security question had been heightened by an unprecedented series of confrontations between, on one hand, oil-consuming countries and the major oil companies, and, on the other, the oil-producing nations over royalty and price increases, government participation, and threats to cut back production. For the short term the U.S. would remain dependent on Middle East oil, but over the long run she would attempt to improve the security of her oil supplies by developing North American resources—conventional oil and gas, synthetic fuels, nuclear power. This policy would have the backing of the oil companies, Levy argued: "As control over reserves becomes less certain in foreign areas and their status as privileged offtakers is eroded, the majors are likely to be increasingly interested in access to North American reserves to meet their North American crude requirements." North American oil will move into North American refineries as a matter of government policy, Levy predicts, thus none of the larger integrated companies can afford to be excluded from reserves such as the oil sands.

Another Levy argument, and one which has evidently impressed Peter Lougheed, is that the oil sands have a competitive edge over other synthetic fuels, oil shale, coal liquefaction, etc. "Pushing ahead now could serve to maintain and perhaps even increase this lead. Alberta probably has the opportunity of developing second-generation plants by the time first-generation U.S. oil shale and coal liquefaction plants are coming onstream. And third-generation tar sands plants, which could be expected to benefit from relatively well established technology and perhaps important cost-savings, might be onstream by the time these other synthetics are in a second-round phase of development. On this basis, the tar sands might be able to maintain a significant competitive advantage in North American markets for some time to come." The oil sands, under Levy's game plan, would be developed in direct competition with the heavily subsidized synthetic fuel projects of the United States.

From the standpoint of both U.S. and Canadian policies, Walter Levy advised the Lougheed government in early 1973, "there is an advantage to moving early and rapidly to develop tar sands production. For the United States, early development of the tar sands could contribute to the availability of secure North American oil supplies over the critical period before its own long-run efforts to develop conventional and synthetic oil might begin to pay off. For Canada, the establishment of early and substantial volumes of tar sands production could be essential to maintenance of Canada's export potential, providing an offset to Canada's rising volume of oil imports." Specifically how rapid the development should be Levy did not spell out; nor does he refer to any of the costs such a strategy would entail—the social, environmental and economic damage which Alberta's civil servants feared a fast rate of development would inevitably bring in its train. Levy's other main recommendation is one he has made in many other contexts: that the major oil companies alone have the technical competence, financial capability and access to markets to properly develop the oil sands. Only the majors meet the criteria of experience, capital and markets: smaller Canadian "independents" could participate in joint ventures with the majors, but only the latter are large enough to lead such projects. Any idea of government ownership and development of the oil sands is never mentioned in the Levy report.

In the spring of 1973, shortly after receiving this package of advice, Peter Lougheed attended an international conference on energy in Bilderberg, Sweden. "There were representatives from seventeen nations," he later told the Alberta legislature in a speech justifying his deal with Syncrude, "such people as the finance minister from Germany, the chairman of the board of Royal Dutch Shell, and Mr. Levy, one of the most informed consultants in the world." Lougheed drew two important conclusions from his talks with Levy and others in Sweden. In the first place, both Europe and the U.S. were making an all-out effort to develop nuclear energy, and the United States was making an equally determined attempt to develop the Colorado oil shales. "The second conclusion I reached from that meeting was that the Alberta oil sands was simply not part of their thinking on the world oil scene, despite its magnitude of potential, because they looked upon the first plant [Great Canadian Oil Sands] as being merely a pilot plant, and they concluded that the technology of even the surface portion had not been established." The government's negotiations with the Syncrude consortium were approaching a crucial stage when Lougheed returned from Sweden, and it is clear from his own account that he had decided that they could not be permitted to fail. "It is critical," he told the legislature, "that Plant No. 2 [Syncrude] get under way. Otherwise we can lose the lead which we now have of perhaps four to five years over the Colorado oil shales. Also nuclear energy could render obsolete, and as a useless asset, the Alberta oil sands and the history of all that is involved." With nightmares such as this in the minds of the government's representatives at the bargaining table, Syncrude could almost dictate its terms and threaten Lougheed's entire vision of the oil sands if their terms were not met.

The cautious, Canada-first advice of Lougheed's civil service, along with its emphasis on the risks and costs of fast, forced development, was clearly regarded by the cabinet as a parochial luxury. While the Alberta government was waiting, carefully protecting the environment and extracting maximum benefits, the oil shales and nuclear energy might pass Alberta by, rendering the oil sands a "useless, obsolete" asset. Walter Levy's gung-ho call for rapid development made even more sense to the government after the Yom Kippur Middle East War in October 1973. The subsequent Arab oil boycott strongly intensified American fears about its oil imports, and OPEC's decision to quadruple posted prices of oil to nearly $12 a barrel suddenly made unconventional reserves like the oil sands seem a highly attractive proposition.

Now Alberta cabinet ministers spoke of the need for "crash programs," holding out the prospect of twenty Syncrude-sized plants by the year 2000. Offers of financial participation poured in from Japan, Europe, the Middle East, even Poland, and visitors by the thousands dropped into remote Fort McMurray during the winter of 1973-74. As companies began to file applications before the Energy Resources Conservation Board in Calgary, Peter Lougheed moved to accelerate the process by announcing his own imitation of Project Independence. Lougheed's "Energy Breakthrough" project, involving provincial government funding of $100 million over five years, would be aimed at developing a workable *in situ* process, through which the deep-seated oil sands and heavy oils could be recovered. (Less than five percent of the oil sands can be recovered through the surface mining techniques employed by GCOS and Syncrude, thus the *in situ* technology is literally the key to bringing the resource into the world energy picture.)

While the Lougheed government moved to implement its own strategy of development, events elsewhere during that autumn of 1973 and the dynamics of Canadian-American energy relations were about to culminate in one of the most bizarre episodes in the tar sands' history. All the details of this story have yet to filter out; but from what is known it is clear that the United States, acting through both informal and official channels, made a strong power play for the oil sands at the height of the energy crisis. Had this ploy succeeded, the sands would now be launched on a path of development the pace of which would make Walter Levy look like a counsel of prudence. Here we must shift our view once again, away from Alberta to Washington and Ottawa where the oil sands were thrust into sudden prominence by the U.S. energy crisis of 1973.

In spite of their huge theoretical potential, the oil sands actually offer very little hope of relief for North America's short-term oil requirements. At the present projected rate of development—one plant every two years initially—the developers will be extremely lucky to reach a mere half million barrels per day output by 1985. By 1990 output from the Athabasca surface mining projects might reach 700,000 b/d, while another 200,000 b/d could perhaps be expected from the heavy oil deposits of Cold Lake and Peace River. This would mean a rather modest total of 900,000 b/d by 1990, a figure which is too low even to meet the projected shortfall in Canadian oil requirements. By 1977 production from the western sedimentary basin will begin to decline at an expected rate of 100,000 barrels a day every year. One oil sands plant per year would be needed simply to meet this Canadian shortfall, leaving no synthetic oil production for export to the U.S.; and no one in Alberta, including Syncrude, is optimistic about constructing plants that quickly. Why then, the question arises, should the Americans be interested in the oil sands?

Even if we turned over all our remaining conventional reserves to the U.S., they would be consumed in under twenty months. The United States in 1973 was already importing 6.3 million barrels of oil every day and that figure, according to some projections, will more than double by 1985. With Canada preparing to cut back her exports of conventional crude when the Interprovincial Pipeline is extended to Montreal and the oil sands coming on stream at a rate too slow to meet Canadian requirements, how can the United States expect significant increased oil supplies from her northern neighbour?

One person who thought he had an answer to that puzzle was Herman Kahn of the Hudson Institute, America's most prestigious "think tank." Kahn, described by Geoffrey Stevens of the *Globe and Mail* as "a huge man who thinks on a huge scale," has long taken an interest in Canadian affairs. Back in the mid 1960s Kahn, author of *On Thermonuclear War, Thinking About the Unthinkable* and *On Escalation,* would periodically drop by the Department of Defence in Ottawa to deliver day-long monologues on various aspects of nuclear strategy and international politics. Academic acquaintances of the writer would stagger home from Kahn's marathon lectures, exhausted and dazzled by his theatrical performances. Near the end of the 1960s Kahn moved away from military strategy, and the Hudson Institute began to spend more of its time proposing solutions to America's social crises, dealing with racial issues, political extremism, riot control and so on. More recently, Kahn has taken to peering into the future, authoring *The Year 2000* and *Things to Come,* and the Hudson Institute has turned to advising large U.S. multinational corporations about how to prosper in an increasingly hostile world environment. Having demonstrated that thermonuclear war can be waged "rationally," and having led flying think-tank sessions in airplanes over Portuguese colonies in Africa or the wilds of Colombia, it was only to be expected that Kahn's social engineering mentality would now come up with a suitably grandiose solution to the industrial world's energy crisis—the Athabasca oil sands.

The bizarre tale began to leak out at the end of 1973. Clair Balfour, a *Financial Times* reporter, revealed that on November 28 Herman Kahn and a Montreal associate, Marie Josée Drouin, had a meeting with Pierre Trudeau, Energy Minister Donald Macdonald and senior members of Macdonald's staff, during which Kahn unfolded his brainstorm for a massive, crash program of development in the oil sands. What Kahn proposed was an internationally financed effort on a giant scale to build twenty Syncrude-sized plants by the end of the 1970s. Together these plants, all using the existing strip mining, hot water extraction and upgrading technologies, would produce between two and three million barrels of synthetic crude daily, most of which would be marked for export in the initial stages. Between 1978 and 1984 some six billion barrels would be siphoned off for foreign consumption; after that Canada would be able to use what oil she required for domestic purposes and would be left with an oil field twice the size of Alaska's Prudhoe Bay. All of this, including full ownership and cheap oil—$3 a barrel in constant 1973 dollars, said Kahn—could be Canada's without investing a dime. The estimated $20 billion investment would be put up by an international consortium of governments—the U.S., Europe and Japan—in return for a commitment of secure oil imports until the mid 1980s. Such government investment would keep interest costs down and the impact on exchange rates would be minimized by having the oil sands isolated from the rest of the Canadian economy. Remarked Clair Balfour, "It would be as though the 10,000 square miles of oil sands were declared international territory, for the international benefit of virtually every nation but Canada." Because labour requirements in such a project would be fantastic, Kahn suggested that a labour force of some 30,000 to 40,000 South Koreans would have to be imported at three to four times their normal wages. "Canadian unions would be bought off by having the imports pay normal dues and contribute to pension plans. They would never collect any benefits."

As we shall see, the companies currently operating or planning operations in the mineable oil sands are encountering severe environmental problems in reclaiming the strip mined surface, in containing atmospheric pollution, and in disposing of used waste water which contains unextracted raw bitumen. These environmental issues are of little concern to the Hudson Institute, "which believes the northern Alberta area to be a relatively undesirable environment anyway. Its restoration would not be a matter of aesthetic quality." One argument Kahn reportedly used to sell his plan was that the oil shales of Colorado were located in such pleasing, mountainous scenery that any crash program would inevitably bring on entrenched resistance from environmentalists; the oil sands could be developed without confronting such opposition. One economist who attended the November 28 meeting remarked, "I suppose if one were at war, it's surprising the things one would do and things you would ignore. . . . If you were at war, you'd use up the Athabasca River and say the hell with it." But who was at war? The Athabasca, it happens, is a major part of the whole Mackenzie River system, which runs from northeastern Alberta all the way to the Arctic. One researcher calculated that under Kahn's scheme some 10,000 to 20,000 tons of unprocessed bitumen would be dumped into the Athabasca every day—enough oily waste to destroy eventually all flora and fauna between Fort McMurray and Tuktoyaktuk!

Kahn's oil sands scheme has generally been dismissed as the grotesque by-product of a somewhat grisly and overactive imagination. Donald Macdonald was quoted by Clair Balfour as being sceptical: "I don't know, within the world community, why we should feel any obligation to rush into such large-scale production, rather than leave it in the ground for future generations." Macdonald was, however, impressed with the amount of thinking Kahn had put into the project and his sensitivity to Canadian nationalist sentiments. The minister added that some sort of trade-off would be demanded to make it appealing, probably in terms of easier access for Canadian manufacturers in U.S., European and Japanese markets. Thus Macdonald appeared to be leaving the door slightly ajar for a deal. A little later, early in January 1974, Mitchell Sharp, then Canadian minister of External Affairs, responded enthusiastically to President Nixon's call for a meeting of the energy-hungry industrialized nations. Sharp added the apparently extraneous note that Canada would welcome more foreign investment in the Athabasca oil sands, so long as control of development remained in Canada's hands.

Kahn, it seems clear, had been acting as an informal envoy for the American government and not as an impartial brainstormer simply trying to pawn an idea. Before appearing in Ottawa he discussed his plan with leading officials of the Japanese and various Western European governments, with Henry Kissinger's staff and with Richard Nixon personally. In June 1974 he also discussed it with the Alberta cabinet. The Hudson Institute has links with the federal Liberal Party, especially with former minister of Supply and Services, Jean-Pierre Goyer. Goyer arranged both the November 28 meeting and an earlier session with the cabinet at which Kahn presented an audio-visual outline of future economic trends for lunchtime entertainment. Goyer was so impressed with Kahn's game plan for the oil sands that he made it his personal mission to propagandize the idea of crash development, outlining modified versions of the Hudson Institute proposal to audiences in Vancouver and Calgary. In Goyer's version the "advantages" to Albertans and Canadians got heavy emphasis. The advantages to Alberta would be fuller development of its resources, high revenues, development of secondary industry and a stable market. Benefits to Canada would be an improved international trading position, eventual energy self-sufficiency and additional income. The minister's efforts on Kahn's behalf finally prompted the *Financial Post* to wonder, "Why is Goyer promoting giant oil sands development?" Could it be, the newspaper asked, "that the government has not rejected completely a fast-development program for the tar sands? Ottawa was earlier under considerable pressure from Washington to develop for the export market whatever petroleum reserves exist in Canada. Japan and Europe presumably would welcome any contribution that Canada can make to world energy supplies." Was Goyer kite-flying a controversial idea to test public reaction for a wavering cabinet?

Behind the scenes, intense diplomatic pressure was building up on Ottawa during the winter of 1973-74 to help out the energy-scarce, industrialized nations. There was, for instance, an offer from Japanese officials for $6 billion investment in a joint Japanese-Canadian petrochemical complex in the oil sands. Japan was offering relatively generous terms—more than fifty percent Canadian ownership—but Ottawa seemed to be more embarrassed than attracted by the proposal. Perhaps the embarrassment derived from some countervailing pressures being exerted by the United States. The Americans are known definitely to have made at least two separate, formal offers to "help" Canada rapidly exploit the oil sands. The first, according to well-placed sources in the federal government, came through the ordinary, ambassadorial channels in the form of an $8 billion package of American industrial investments. Central focus of the offer was a crash program for the oil sands, and as a sweetener some additional investments. The proposal was given cabinet consideration but discussion became bogged down over concern for the impact on the Canadian dollar and the inflationary effects of so much rapid capital expenditure. The second known U.S. offer came at the end of January 1974 during discussions in Washington between Donald Macdonald and former U.S. energy czar William Simon. The day before his talks with the Canadians, Simon publicly linked the oil sands to the proposed Mackenzie Valley gas pipeline, a project the Liberal government was doing its utmost to support. The U.S. delegation urged Macdonald to consider broadly based energy commitments, including arrangements for long-term export supply and a massive, multi-billion dollar acceleration of oil sands development. Macdonald was reported in the press as having rebuffed these overtures—"We cooled their ardour on it," he remarked—and having told the Americans to forget about Canada as a guaranteed long-term supplier of oil and gas. The strains on the Canadian economy produced by accelerated oil sands exploitation would not serve Canada's interests; in any event, the oil sands were needed to satisfy domestic oil requirements. Only large discoveries of offshore or Arctic oil could alter Canadian plans to phase out its oil exports to the U.S.

Apparently Macdonald, to his credit, had not acquiesced in the U.S. oil sands power play, but it cannot be entirely ruled out that his stance in Washington was really a bargaining posture. American pressures on Canada to enter into formal continental energy-sharing arrangements go back at least to 1970 and have been intensified since the Arab oil embargo in late 1973. In spite of rumours about the "blue-eyed Arabs" withholding energy supplies from an ally ("kicking a friend when he's down," moaned John Poyen of the Canadian Petroleum Association when Ottawa boosted its oil export tax), Canada actually increased her oil exports to the U.S. by twenty-one percent in 1973. Production from oil fields in Alberta reached an all time high, straining the capacity of the producing wells and pipelines to the utmost. In November Pierre Trudeau announced that the Interprovincial Pipeline would be extended from Sarnia to Montreal, holding out the possibility of Canada achieving self-sufficiency in oil by decade's end. Yet in spite of the fact that this step would greatly reduce the amount of oil surplus to domestic requirements and thus available for export, Trudeau nevertheless promised that oil exports would be kept up for years to come. Most authorities strongly doubt that Canada can regain net self-sufficiency in oil during the 1980s, so serious is the looming domestic shortage. But the prime minister did not share his secret plan for squaring the circle.

Would the government of Canada do the unthinkable and turn the oil sands over to the energy-hungry industrialized nations for a wartime program of crash development à la Herman Kahn's scenario? This may seem far-fetched, especially since the promised return of less than three million barrels of oil a day seems hardly worth the massive costs. But the possibility of a new oil sands power play cannot be dismissed entirely. Were the OPEC nations, for instance, to renew their oil embargo against the West or make substantial cuts in their overall production, then the pressure for accelerated development of oil substitutes would become extreme. In February of 1974, the major oil consuming countries, with France dissenting, agreed in Washington on the need for a comprehensive "action program" to deal with all facets of the world energy situation by cooperative measures. The action plan includes "the sharing of means and efforts while concerting national policies" in such areas as: conservation of energy; emergency allocation of oil supplies; acceleration of development of additional energy sources; and acceleration of energy research and development programs. The script for the Washington energy conference could have been written by Walter Levy himself. Canada's participation at the Washington conference and in the International Energy Agency set up to implement the action program means that the oil sands will inevitably come under some pressure for accelerated development. In a fresh crisis between consuming and producing nations contingency plans could quickly become reality, particularly since the Canadian government contains many key officials who believe that the country's role in defending the "Free World" involves sharing our resources with our allies at bargain basement prices.

At the very least, it can be confidently predicted that the U.S. government will continue to exert strong bilateral pressure on Ottawa for increased energy supplies and the rapid exploitation of the oil sands and Arctic resources. The United States needs guaranteed, long-term commitments of energy and there are several bargaining areas where she can be expected to exert maximum pressure. First, the Liberals badly want the Mackenzie Valley natural gas pipeline, and to get it they are willing to negotiate a mutual security treaty with the U.S.; the Americans can be expected to attempt to link this question to long-term oil supplies from the oil sands. Second, as Donald Macdonald has hinted on several occasions, Canada wants easier access for her manufactured products in the U.S. market; a trade-off of energy for guaranteed industrial markets has often been suggested from the American side. Third, there is the Auto Pact, a continental arrangement in a key sector of the Canadian economy which gives the Americans a vital pressure point in their dealings with Ottawa. And finally, the American government can pressure American-owned oil companies and indirectly blackmail federal and provincial authorities worried about future investments. If Canada takes too strong a line on future exports and refuses to enter into guaranteed commitments of future energy resources with the U.S., then Washington might well try to persuade American energy companies to make their heavy investments in the oil shales, nuclear energy and other domestic resources instead of uncertain reserves such as the oil sands. The U.S. government may well argue that American monies should not be invested in the oil sands unless the investment yields something in return by way of secure energy imports.[2] This threat, hinted at by U.S. officials on various occasions, probably explains why Canadian politicians continue to promise exports when they know their own country will soon be a net importer of oil.

It was appropriate that the newly appointed American ambassador to Ottawa, William J. Porter, should choose Project Independence as the theme of his first public offering to Canadians. "Because no one is closer to us," the ambassador told the Men's Canadian Club of Ottawa on May 9, 1974, "we think immediately of Canada when we discuss the scope and importance of the president's announcement of Project Independence." While Americans got their drive for self-sufficiency under way, they would hope to continue their traditional cooperation with Canada. Came the implied threat: "We have noted the views of those in Canada who urge that you hoard your resources and that you end your petroleum shipments to us now. You will make your own decision about that." The transfer of energy material was not one-sided, the ambassador noted. The U.S. exports coal to Canada "and of course the interconnection of our electric power grids at a number of points is of critical regional significance to utilities on both sides of the border. I feel confident personally that you will take into account the advantages and disadvantages of various courses of action you may consider for the short, medium and long term ahead." Then came the carrot, the chance to cash in: "I see no reason why the same spirit of enterprise that we have jointly carried through successfully and which brought such benefit to us should not be repeated. We have much to offer each other, in the new era ahead."

Canada could continue to enjoy the good life, Mr. Porter was saying, but only in a new era of revitalized continentalism.

6   Oil Insurance

"The American Beauty rose can be produced in the splendour and fragrance which bring cheer to its beholder only by sacrificing the early buds which grow up around it. This is not an evil tendency in business. It is merely the working out of a law of nature and a law of God."

John D. Rockefeller, Jr.

During the winter of 1973-74, the American petroleum industry geared up for the most ambitious and extravagant advertising campaign ever undertaken in that ad-soaked society. The central message of the campaign was that the large oil companies were innocent of the many charges that they had "rigged" the energy crisis. Millions of dollars, many of them tax deductible, were poured into television and radio announcements, newsletters, full page newspaper ads and speakers bureaus (Gulf Oil alone employs some 350 people in its "Vital Source" speakers bureau). The U.S. media have been saturated with the now monotonous line that the era of cheap energy has passed, that high profits and higher prices are needed to develop new energy resources, and that the shortages and lineups at the gas pumps are the inevitable result of naive and irresponsible meddling with the free enterprise system by politicians and environmentalists. Suddenly, the world's champions of conspicuous energy consumption had turned into concerned conservationists, urging people to turn down their thermostats and go for a walk instead of a drive. "The sooner we face the facts, the better," warned Gulf sternly.

But evidently the "facts" have not gotten through. In spite of, or perhaps because of, this unprecedented advertising blitz—the huge costs of which would ultimately be borne by taxpayers and consumers—Continental Oil's public relations chief subsequently confessed that people "will not read or believe anything we say about energy or profits." Announcing the largest annual profit ever made by a single corporation—$2.4 billion for 1973—at an unprecedented news conference, Exxon's former chairman, J. K. Jamieson, glumly acknowledged that the oil industry's image had never been worse. The polls bear him out. Upwards of eighty percent of those Americans polled believed that the ads were "phony" and that the oil companies had manufactured shortages to raise their profits. Exclaimed one angry businessman after storming out on a speech by the consumer representative from Mobil Oil in April 1974, "That's like the Mafia trying to tell you they're nice guys." After being assailed for over an hour by angry shouts and derisive laughter from the Staten Island Kiwanis Club, Mobil's representative pronounced her audience "a very moderate group." Earlier in 1974 she spoke to the Sierra Club and found herself "mentally exhausted with them. Their preset attitude was so heavy that I really would have had to streak across the room to get them to listen halfway."

At the root of the oil industry's credibility problem lie the astronomical profits which the biggest companies have made from energy scarcity in the United States, a scarcity which many Americans believe was contrived. It is the issue of profits, what one writer calls "a financial feast in the midst of fuel famine," which above all else keeps the ad agencies churning out the copy. The combined net earnings of the largest international companies, the so-called "seven sisters"—Exxon, Shell, Mobil, Gulf, Texaco, Socal and British Petroleum—stood at $9.15 billion in 1973 and $11.65 billion in 1974, a true financial feast unparalleled in the history of the modern corporation. The feast was so conspicuous, however, that it forced the oil industry to explain itself. In 1974 a group of leading U.S. oil executives, raised on the philosophy that making profits made America great, were asked by investigating senators whether they were not "ashamed" to be making so much money. The industry's reply has been that the profit bonanza was actually illusory—the profits were mostly "inventory" profits. Accounting procedures were revised to show the "true" earnings of the companies and these were uniformly lower than the so-called inventory profits. One day someone who can combine inside knowledge of the oil business with the spirit of the theatre of the absurd will write a book on the elusive, now you see it, now you don't character of oil profits. As we shall see in our later exploration of Syncrude, the question "what is a profit?" is a central one in the politics of oil.

Unhappily for the oil companies, their success in achieving high profits is not representative of a general condition of prosperity in the United States. Quite the contrary, the U.S. entered 1974 in a rather dismal business slump as the major auto producers and many other leading manufacturers, feeling the pinch of high energy costs and the psychological uncertainty of investors and consumers, cut back production and laid off thousands of workers. In sharp contrast to oil's escalating profits, one of the world's biggest corporations, General Motors, suffered a colossal eighty-five percent drop in first quarter earnings over its 1973 level. By 1975 the big question about the American economy was whether it was stalled in a temporary recession or heading for a full-scale depression. It is an ill moment for any industry, but particularly one with the ruthless traditions of oil, to be making record profits. Consequently, and not for the first time, the American petroleum industry has been in a good deal of political hot water.

Arising out of the pronounced suspicions of ordinary working Americans, a series of investigations of the oil industry have been launched in the United States by federal regulatory bodies, various state governments and citizens' lobbies. Together with the fruits of earlier inquiries undertaken in similar circumstances, these have managed to pierce the secrecy of the industry and have yielded some intriguing clues about the inner workings of this most concentrated system of private power on earth. It is in the structure and motivations of the international oil industry that we can locate the origin of many of the pressures being exerted on the Athabasca tar sands and on other North American energy resources. For these areas have become the objects of a power play by the oil industry in its bid to maintain its dominance over the present world petroleum economy and to corner the market on the globe's future energy fuels.

Naturally enough, the large oil companies hotly deny that they are engaged in any international struggle for power or that their far-flung operations around the globe take them beyond strictly economic activity. Oil's spokesmen go to great pains to stress that their sole purpose is the commercial production and marketing of vital commodities for an energy-thirsty world; what motivates the oil industry is the pursuit of profit, not power. And with equal indignation they protest any suggestion that the industry's behaviour can be described as monopolistic or oligopolistic. The oil industry, we are informed, is one of the most competitive businesses in the world. The public utterances of oil is replete with the rhetoric of laissez-faire capitalism, lauding the benevolence of competition, the market place and profit making while denouncing capricious bureaucrats who interfere with the free enterprise system. It should therefore come as no real surprise to learn that the oil industry's solution to the energy crisis is simply to permit "the market place" to straighten out the mess by allowing prices and profits to rise.

A mass of evidence flatly contradicts these public postures. The oil industry must be judged by what it does, not by what it says. The evidence suggests, first, that the international petroleum industry has never been ignorant about the realities of power nor timid about putting its power to effective political purposes. Where the control of vital resources and so much economic wealth is concentrated in a very few private hands, fine distinctions between what is economic activity and what is political activity become rather academic. The nature of the oil business guarantees that its activities will have a broad social impact; it cannot be considered a private concern simply because what it does affects the lives and livelihood of so many. Professor E. H. Shaffer, author of a study of American oil policy, reminds us: "The possession of energy resources implies something more than the ability to feed an industrial complex. It implies a fundamental power relationship. Those who have effective control over those resources, be they private companies or national governments, have power over those who do not. In a world based on unequal power relationships, the possession of energy is a *sine qua non* for maintaining one's relative position." Students of Canadian-American relations would do well to keep Shaffer's point in mind: where one country and its corporations effectively monopolize another's energy resources and the ways in which they are developed and utilized, then that country also has the leverage to shape many of the choices and alternatives supposedly open to the dependent nation. Energy is not an abstraction; it is bound up in the distribution of social power within and among nations—a fact which should be immediately apparent to anyone familiar with power relationships between the United States and Canada.

Oil is the supremely political commodity. Robert Engler, author of the classic study, *The Politics of Oil,* remarks that oil "offers an extraordinary case study of how economic power becomes political and social power." It is Engler's thesis that the oil industry derives much of its power and influence from the "vacuum of public policy" it finds all over the world. The influence of oil, in other words, depends on the unwillingness of government to assert its legitimate jurisdiction over energy affairs; in abdicating responsibility to a socially irresponsible system of power, public government in effect creates the "private government of oil." In the absence of an energetic and watchful public policy, "the highly developed private government of oil seeks the support of public government wherever its own political and economic machinery is inadequate for fulfilling industry objectives."

Oil men still portray themselves as self-reliant and they love to romanticize about an ideal society of rugged individualists. But the truth of the matter is that no business is more dependent on the helping hand of government than their own. It is fascinating to speculate about what might happen to these great champions of free enterprise, who spend so much time urging government to keep out of the affairs of the oil industry, if government decided to do just that. In Canada all one has to do to rattle the cages of Calgary's dinosaurs is to propose that the oil industry should henceforth be taxed at the average rate of Canadian manufacturers. Without the various publicly supported props and supports it has managed to have legislated over the years—the depletion allowances, the tax deductions on foreign operations, the "conservation" boards, the import quotas, the immunities from antitrust action, and the diplomatic and military backing of the world's most powerful nation—it is conceivable that the whole intricate, interlocking system that is the world's oil business would have long ago disintegrated. Syncrude's heaviest costs and risks are being underwritten by the Canadian and Albertan governments, but this model of free enterprise in action is not exceptional. Oil, quite simply, is the biggest and most profitable welfare case on earth.

Even more cynical than the alleged political virginity of the oil industry is its commitment to what it calls the forces of the free market. Here once again, the very last thing the petroleum business wants to see introduced into its affairs is actual competition. In many ways, the history of the oil industry is the history of a long struggle against the threat of outside competitors and internal rivalry. There is, in point of fact, no "free market" of oil, the laws of supply and demand have remarkably little to do with pricing or any other aspect of the industry's operations. The major oil companies, to quote a recent study by the U.S. Federal Trade Commission, "continually engage in common courses of action for their common benefit." Although many oil executives still pay lip-service to Adam Smith's vision of a shopkeeper society, in their hearts they believe far more in the dictum of John D. Rockefeller, Sr., founder of the old Standard Oil trust: "The day of combination is here to stay. Individualism has gone, never to return."

It was this underlying sense of realism and mutual interest which brought together the heads of what were then the big three of world oil—Royal Dutch/Shell, Jersey Standard (now Exxon) and Anglo-Iranian (now British Petroleum)—in the civilized surroundings of Achnacarry castle in Scotland one autumn weekend in 1928 to draft the foundations of the international oil cartel. Ostensibly there to hunt grouse, the oil executives had actually convened to head off a destructive price war and a crisis of overproduction and oversupply, the nemesis of the world oil market. "Up to the present," the grouse hunters explained, "each large unit has tried to take care of its own overproduction and tried to increase its sales at the expense of someone else. The effect has been destructive rather than constructive competition resulting in much higher operating costs." Accordingly, in the interests of "constructive competition" a set of "as is" agreements were worked out to divide and stabilize the world market, to reduce surplus production and supply, to eliminate price competition and other competitive practices which increased costs, and to exclude potential interlopers. It was only a beginning, but by spelling out the areas of mutual self-interest linking the largest oil companies and implying that competitive practices were destructive and irrational, the Achnacarry gathering helped establish the ground rules of the international petroleum cartel. One can but wonder what Adam Smith would have made of the grouse hunters and their latter-day successors.

Much light was thrown on the structure and internal workings of the international oil industry in the aftermath of the Achnacarry agreement by the U.S. Federal Trade Commission during one of the FTC's unsuccessful anti-trust campaigns against the big oil companies. The publication of the FTC's ambitious and extensively documented 1952 report, *The International Petroleum Cartel,* illustrated in great detail how seven giant Anglo-American companies—Jersey, Gulf, Texaco, Mobil, Standard of California, Shell and British Petroleum—had managed to work out and police the most elaborate and complex cartel agreements on pricing, markets, production and so forth. Outside the U.S. and the Soviet bloc, the seven sisters controlled in 1949 ninety percent of the world's crude oil reserves, eighty-eight percent of production, seventy-seven percent of refining capacity, eighty-five percent of cracking capacity (that part of the upgrading process that transforms crude to various products and feedstocks), two-thirds of the world's tankers and virtually every pipeline in existence. From oil well to gas pump, at some point along the line any smaller company was forced to buy from, sell to, or use the facilities of the vertically integrated majors. Moreover, the big seven complemented their own vertical control of all levels of the oil business by engaging in interlocking horizontal joint ventures with each other in exploring for, producing and transporting oil all over the world. Joint ownership assured control over foreign reserves, but it also offered the big companies the occasion to coordinate their policies toward several producing countries simultaneously. "The directors of Standard of New Jersey [Exxon] and Socony-Vacuum [Mobil], who determine the policies of the Arabian-American Oil Co. [of Saudi Arabia], are the same men who help shape the behavior of the Iraq Petroleum Co. The directors of the Anglo-Iranian Co. [B.P.], who assist in making high oil policy for Iraq and Iran, participate, along with the directors of Gulf, in planning the price and production policies in Kuwait." The boards which managed the myriad joint companies could function as private planning committees where differences could be resolved and an "oil policy for the world" established.

Historically, the international oil companies have maintained their predominance through the relentless pursuit of several key strategies. Like all good oligopolists, they try to control output in order to maintain common price levels which will maximize their profits—they avoid price competition—and they have a mutual interest in excluding outside competitors, whether such interlopers be aggressive smaller companies or national governments. On the other hand, the cartel companies have been in conflict over concession rights, and they are secretive about new discoveries and technology. Another Federal Trade Commission report on the industry, this one dated July 1973, argues that the behaviour of the eight largest U.S. oil companies "should properly be regarded as cooperative, rather than competitive, with respect to: influencing legislation; bidding for crude leases; establishing the purchase price of crude oil; transporting crude oil; refining crude oil; marketing gasoline." On the many levels at which they interrelate, the FTC argues, "the majors demonstrate a clear preference for avoiding competition through mutual cooperation and the use of exclusionary practices. Together they dictate a common price for raw material and seek to stabilize a price for refined product. Their common conduct with respect to pipelines and their tendency to bypass the market mechanism through the use of exchange and processing arrangements, has been clearly exclusionary. These exclusionary practices are directed at a common target—the independent sector of the industry." The majors, adds the FTC, behave "in a similar fashion as would a classical monopo-

list: they have attempted to increase profits by restricting output.''
The spirit of Achnacarry lives on.

If there is any one strategy which will guarantee market power and also exclude interlopers, it lies in control over the supply of the basic raw material, crude oil. Where a small group of companies is able to control supply of the necessary raw material, it becomes so difficult for competitors to break into the industry that monopoly market power and hence monopolistic prices and profits are virtually guaranteed. It was this simple, fundamental truth which eventually led the major companies to orient their development strategies around the principle of self-sufficiency within their organizations for the production, refining, transportation and marketing of oil. And it was the same quest for supply of the basic raw material which impelled the majors into the international hunt for vast concessionary rights in the Middle East, Venezuela, Russia and elsewhere. A company can build refineries, buy tankers, hire geologists and set up a marketing network—but without a secure supply of crude oil all of this is quite irrelevant. Obviously, then, the few companies who can build up the reserves to control the bulk of supply are strategically positioned to charge excessive prices for crude oil and thereby wipe out profit margins for independent refiners. And further, any smaller company which has ambitions to break into the fabulously profitable but closed industry must quickly capture its own source of crude supply or face an uncertain future of dependence on the majors. Well aware of this and of the fact that the globe contains a great deal of known and undiscovered oil. the majors developed a common horror of upstart competitors gaining access to large reserves of cheap oil and using such oil to threaten the stability of high prices. Actually, most of the smaller companies know a good thing when they see it and have been content to play by the rules of the game as junior partners in the cartel. But the senior partners nonetheless perceive themselves as well-established landowners who constantly have to fight off interlopers, and their strategies should consequently be understood as partly a defensive reaction to this threat, as an attempt to hold and expand their domain against the challenge of new competitors. As an insurance policy against such threats, the integrated majors have hedged their risks by investing in exploration ventures and tying up concessionary rights in new areas all around the world. "Even if they seemed already to have extensive reserves in the Middle East and elsewhere,'' remarked one observer of the postwar scramble for new reserves, "the fundamental insurance principles of spreading the geographical risk and matching the competitor's locations continued to operate.'' They still do.

The integrated majors have also learned from hard experience that it is bad business strategy to have all of their resources committed in a particular producing nation. This would leave them far too vulnerable to nationalization and political pressure. They have learned that it is prudent to have a well-diversified supply of reserves as insurance against the threat of expropriation and as leverage for bargaining with producing countries. The cartel companies won a bitter struggle against the Mossadegh regime in Iran in the early 1950s because they were able to increase production in countries like Venezuela and thereby isolate Iran from the world oil market. Surplus capacity based on a well-diversified reserve strength permits the companies to play off one producing area against the others, provided the producing areas are not themselves organized into a solid bargaining unit. Exxon, by far the wealthiest and largest of the majors, began more than a decade ago to reduce its dependence on Middle East oil by launching an extensive exploration campaign for new reserves elsewhere in the world. This strategy has evidently paid off handsomely in the company's acquisition of major new holdings in politically stable areas. Exxon controls reserves of some 50 billion barrels of crude oil and natural gas liquids, and surplus capacity of this kind provides the corporation with enough strength to survive several nationalizations—a tremendous asset when it comes to political bargaining with various producers.

The ''seven sisters'' have built their empires around the concept of planned scarcity of crude oil supplies. What the majors have feared above all is the emergence of an uncontrolled abundance of cheap oil or oil substitutes and the upsetting of all of their carefully planned arrangements for keeping supply in balance with global demand. Now clearly, large-scale production of synthetic fuels from the world's immense untapped reserves of tar sands and oil shales or from the transformation of coal into oil and gas pose just such a threat to planned scarcity. We have seen that development of the Athabasca tar sands was held up for nearly two decades in spite of the fact that the technical problems had been resolved by government researchers and that commercial production at prices competitive with crude oil was regarded as feasible in 1950. Similarly, the U.S. Bureau of Mines began research in the Colorado oil shale deposits in 1916 and this work, like that of Karl Clark and the Alberta Research Council, led to the operation of a pilot project in the 1920s. Later, shortage fears during World War Two led Congress to authorize funding for new research into synthetic fuels and the Bureau of Mines opened another experimental plant at Rifle, Colorado. Work at Rifle went on from 1944 to 1955 and confirmed speculation that recovery was technically feasible and that production on a large scale might make oil from shale competitive with conventional crude. It is not surprising then that the major oil companies came to regard the Rifle plant as a potential threat to their overriding interest in restricting world supply. Already in the early postwar years they had been under severe pressure from a host of smaller, aggressive U.S. companies looking for supplies of cheap overseas crude. The majors therefore launched a campaign to get the U.S. government out of the oil shale business and to strangle the pilot plant at Rifle. One of the earliest decisions of the Eisenhower administration when it came to power in 1953 was to terminate funding for the Rifle plant; no further government work would be carried on, henceforth the development of the oil shales would be entrusted to the oil industry. It was almost precisely at the same time that the Alberta government closed down its plant at Bitumount and placed the tar sands under lease to the same companies.

The oil shales and tar sands have always been regarded by the integrated majors as sound oil insurance. In the wrong hands and developed too early, such reserves could pose a threat to world production and pricing arrangements; but in the hands of the same companies which controlled access to most of the world's abundant, cheap crude, the shales and tar sands could be kept out of major production until future circumstances dictated such development. For the oil companies have always understood that they could lose control of the supply of crude oil and that one day the age of conventional oil and gas would be over. A former chief economist with Exxon explains that, "The oil industry wants to avoid the mistakes of the railroads. The railroads thought their business was only hauling passengers and freight by rail. They forgot there were many other ways of doing this, such as airplanes. They weren't smart enough to realize they were really in the transportation business, and now they're close to bankruptcy."

Faced with this unsettling precedent, the oil companies have been looking at the future energy market and moving aggressively into other basic fuels. They anticipate a continuation of the trend toward increased fuel competition, particularly in areas of the U.S. economy such as the electric utility sector. Here both coal and uranium are already roughly competitive with oil and natural gas, and in addition coal transformed into oil will be able to penetrate the transportation sector of the economy, the traditional exclusive preserve of the petroleum companies. The general movement toward a common economic market for the four primary fuels—oil, natural gas, coal, and uranium—and the prospect of a large synthetic fuels industry pose significant problems for the oil industry. Predictably, its response has been fitted to the objective of maintaining its predominance in the energy market. Through takeovers of established firms and the creation of new divisions within their corporate systems, the oil companies are now well-embarked on a strategy which will see them ripen into full-blown energy companies. A recent study of concentration in the U.S. energy market comments drily: "The general pattern of interfuel diversification displays a movement of petroleum producers into coal and uranium production. Control over supply in the latter two markets can have an important influence on final product prices as well as on the pace of technological change." In other words, the oil companies are in the process of taking over America's other energy businesses, particularly uranium and coal, and this eventually will give them the power to control prices and production across the entire energy market. Interestingly, what replaces the fossil fuel economy will ultimately be decided by the same few firms which now control that economy. Naturally enough, the policy of diversification into other fuels has spilled over into Canada: the branch plant oil industry is using its high Canadian returns to corner the market on our future energy needs. Whether Canada gets gas from the Arctic, oil from tar sands or nuclear power from breeder reactors—or from all of these—depends to a significant extent on the investment decisions of the American petroleum industry.

Contrary to company and media information, the move into the future fuels, including the tar sands, has not been prompted by a sudden world shortage of oil. The "energy crisis" is not the result of a physical shortage, it is policy induced: the energy crisis is a political crisis. Far from running out of conventional oil, the summer of 1974 saw the world return to normal surplus conditions, with supplies outstripping demand by about two million barrels a day. There are, assuredly, individual countries which face the prospect of growing dependence on imports; both Canada and the U.S. fall into this category, but it is simply not the case that the world has almost overnight moved from a state of overabundance to one of physical scarcity. For the short- to mid-term, the world will remain largely in a fossil fuel economy; for the near future, crude oil will continue to be physically in a state of surplus. What is now very much at stake, however, are the political and economic circumstances which will determine how much of this surplus crude oil will be made available and at what price.

What impels the oil companies toward development of the tar sands now is a combination of economic and political interests. Much higher oil prices are obviously one incentive to get into production as quickly as possible. Some of the smaller companies, companies like Sun, Continental, Phillips and Cities Service, may be either net purchasers of crude or are deficient in North American crude oil production and are therefore looking to improve their supply positions and their integration ratios. The majors are better balanced in North American oil supplies, but undoubtedly they are interested in improving their position because of the trend in U.S. oil policy toward a "fortress North America" energy stance. Oil consultant Walter J. Levy explains the corporate implications of Project Independence this way: "The majors are probably convinced that governmental policies over the long run will be designed to ensure access to markets for indigenous resources that are developed and at prices which will permit reasonable returns, given the risks involved. North American oil will move into North American refineries—as a matter of governmental policy if not of free market economics. If the majors have to process North American oil in their facilities, they would rather be the producers of that oil than have to purchase it from others. For the integrated company, owned oil is almost always preferable to purchased oil, because of the margin of profit and cash flow from production. Furthermore, a major company can be at a considerable disadvantage in product markets if it has to compete on the basis of purchased oil against competitors who have access to cost oil. Even worse is the prospect of competing in product markets on the basis of oil purchased from competitors." The drive in American energy policy for secure and stable supplies, in other words, holds important economic implications for the big oil companies: the corporate strategy is to get into North American supply now to avoid the future prospect of having to refine someone else's oil.

But this is not all. The majors are involved today in a new and complicated poker game with the major oil-producing nations, the OPEC bloc, who control fully eighty-five percent of the world's oil. Founded in the late 1950s in reaction to declining prices and the cartel's worldwide manipulation of producers, OPEC has only recently been able to challenge the supremacy of the majors and to effect a redistribution of power in the global politics of energy. Up to about

1969 the majors still held effective control over price, production levels, investments and so on, and in addition they enjoyed a strong bargaining position with the key producing countries. Since 1969 a good deal of this power has been cut away by OPEC. Although the companies are making higher profits than ever before from their overseas operations, they have also lost a good deal of their monopoly power over prices and production because of OPEC's new assertiveness. What has been happening in countries like Venezuela, Saudi Arabia and Iran is that the affiliates of the international companies have been losing some of their exclusive concessionary privileges in return for a new and admittedly lucrative role as tax collectors for OPEC and contractors providing technical services. Through so-called "participation" agreements, some oil-producing nations have purchased partial or total ownership of the affiliates of the international companies. But participation does not amount to nationalization; rather, it is an ingenious alternative to nationalization which permits the companies to have continued privileged access to crude supply—they buy back the government's share—and keeps the oil producers themselves out of the downstream operations of the oil industry. Saudi Arabia's oil minister has argued that nationalization would be the end of OPEC's solidarity, since each country would then be responsible for selling its own crude in the world market—a situation conducive to competition rather than cooperation. The producers thus have an interest in keeping oil companies as crude oil marketers, while the companies in return retain access to oil and make a handsome return for servicing OPEC. The companies and OPEC also have a common interest in high prices—the companies pay their taxes, then pass them on to consumers for better profits—and a common interest in controlling production.

The problem, as the majors must see it, is that they have surrendered a good deal of their power over price and production to OPEC in return for high short run gains and long-term insecurity. The relations between the big companies and OPEC are unlikely to be stable over the long run, indeed OPEC itself may turn out to be unstable, and the companies face an unpredictable future in the world's major producing regions, in particular the Middle East which controls some sixty percent of world production and continues to be the pivot point in the international market. The oil companies have no intention of simply abdicating their traditional interest in the world's richest oil deposits, and they will adapt themselves to any number of new roles in order to keep a foothold. But the point is that their power has been significantly reduced and their tenure as privileged buyers cannot be regarded as secure. And this poses a challenge to their long-term corporate strategy.

What is threatened by OPEC's assertiveness is the interest of the big companies in a strong resource base to support their downstream investments. To repeat, historically the key to market power has been in secure access to crude oil supply: without this a company faces an unpredictable future. Middle East oil cannot now be regarded as secure; the companies find their privileged positions and their concessionary rights in some jeopardy. This gives the majors an overriding incentive to build up new capacity in more politically secure areas of the world, such as the North Sea, Alaska, or the rest of North America. Such surplus capacity also strengthens the bargaining position of the companies in their future dealings with OPEC and other producers. The large companies are consequently reinvesting some of today's profits in tomorrow's resources, including oil shales, tar sands, synthetic fuels from coal, and spreading their geographical risks.

And so the oil companies now find their prudent insurance policies paying dividends. After sitting on the tar sands and any number of other undeveloped "future fuels" for some time, they may now find it in their economic and political interest to develop these resources in order to improve their supply position and to strengthen their overall bargaining power. Yet the building and maintenance of new surplus capacity on a worldwide basis is extremely costly, involving heavy immediate capital investments and the prospect of these investments showing returns only after a considerable time lag. The energy companies therefore intend to see to it that the development of new energy resources is heavily subsidized by government and that many of their risks are underwritten by the public.

In addition, the oil companies are demanding much higher overall returns today so that they can finance their share of new investment out of retained earnings and thus avoid increased dependence on outside capital markets. The success of this strategy in Canada is reflected in our rising oil and gas prices, and in the corporate profits from the production of western Canadian oil and gas between 1972 and 1975—profits which, in spite of an increasing take by both federal and provincial governments, have reached truly astronomical levels. Between 1972 and 1974, when crude oil went from $2.85 to $6.50 a barrel and natural gas from 18¢ to 60¢ per thousand cubic feet, the gross value for the Canadian inventory of

7.7 billion barrels of oil and 52.5 trillion cubic feet of gas was increased by $50 billion. This should almost double again if Ottawa's intention to raise Canadian oil prices to world levels and to tie gas prices to the concept of "full commodity value" is rapidly realized. Such increases are justified on the grounds that higher prices and better returns are required to bring forth new supplies. Three years have passed since prices began to climb; oil profits for all the major producing companies have increased by 176 percent (Imperial), 310 percent (Gulf), 124 percent (Texaco), 119 percent (Shell), 381 percent (Husky), and so on. The oil industry, taken as a whole, is increasing its profits far more rapidly than any other major sector of Canada's economy. Yet, while oil profits and cash flow have dramatically increased, exploration spending by the industry has risen only gradually for Canada as a whole, and for Alberta and Saskatchewan it has fallen slightly! Canada's inventories of oil and gas are declining while prices and profits have increased enormously, but our governments continue to buy the oil lobby's refrain about letting "the market" solve our scarcity dilemma.

The companies' conditions for development of the tar sands include much higher prices and returns, but they also want to transfer much of the risk and the heaviest costs involved to the public sector. Thus government must shoulder the enormous financial burden of building the massive infrastructure required to service and supply these remote projects, provide equity and debt financing, royalty holidays, guaranteed returns and prices, ensure labour stability, train a work force, underwrite environmental studies and costs—all of which carries a price tag in the billions. In addition, the companies have an interest in seeing both levels of government participate in these ventures, albeit with a minority voice, since this provides the private partners with potential leverage over future energy policies and gives the governments a vested interest in higher prices. Free enterprise is dead, long live the joint venture!

Those companies best positioned to impose their will in the tar sands include most of the top fifteen or so international petroleum firms. The leading lease holders are Imperial, Shell, Texaco, Cities Service, Mobil, Standard of Indiana, Petrofina, Gulf, Chevron, British Petroleum and a few others. This handful of corporations hold leasing privileges throughout the tar sands and heavy oil deposits of Alberta, but they have an absolute stranglehold on the prime mineable area along the Athabasca. Imperial is the most active of the lease holders in the Cold Lake area while Shell dominates in the Peace River. These two giant companies—Exxon Corporation of

New York and Royal/Dutch Shell—should probably be considered the key pacesetters in the tar sands. Where these two lead others will follow—a status which reflects both their stated priorities and their acknowledged seniority in the pecking order of world oil.

We shall shortly return to the power of these companies to dictate their conditions for resource development. First, however, we must assess some of the stakes involved and attempt to weigh the costs and benefits of what is now unfolding along the Athabasca River. What are the costs of forced growth?

7   Forcing Growth

'On one hand we can continue the policies of the conventional crude oil developments creating tremendous and unregulated growth and developments resulting in short-term benefits accruing to the province as well as the long-term costs arising from exported energy, technology, job opportunities and environmental damages in addition to the depletion of nonrenewable resources.

"Conversely, we can regulate the orderly growth and development of the bituminous tar sands for the ultimate benefit of Alberta and Canada in order that Canadian technology will be expanded, Albertans will find beneficial and satisfying employment within its diversified economy, and our environment will be protected and enhanced for future use. But when the magnitude of the real, fiscal and manpower requirements and the environmental consequences are visualized, it becomes apparent that the latter course of action is imperative."

"Fort McMurray Athabasca Tar Sands Development Strategy,"
Conservation and Utilization Committee of the Alberta Government,
August 1972

Contemporary Alberta sometimes seems obsessed with growth. If the rest of Canada is still beguiled by, but also more than a little cynical about the flawed promise of limitless economic development, among many Albertans the doctrine of open-ended progress through individual initiative is virtually an article of faith. Albertans are generally understood by other Canadians to be a somewhat conservative breed, yet conservatives by definition are sceptical of the notion of progress and its alleged rewards. A true conservative politi-

97

cian would be unlikely to close a major pronouncement, as Peter Lougheed ended his Syncrude speech, with rhetorical evocations of a bright future and a canned rendition of "With Our Eyes Upon Tomorrow." What is striking about the "western alienation" ideology of the Alberta government is not the typical preoccupations of conservatives with preserving past traditions, rolling back the encroaching bureaucracy of government, and avoiding the social instability of unregulated industrial expansion. On the contrary, Alberta's dominant ideological outlook is characterized by an upbeat emphasis on change and the future, a pronounced tendency toward intervention in the economy, expanding bureaucracy and budgets, and a fixed belief in the absolute necessity for industrialization.

The fact is that many real conservatives have become increasingly suspicious of the Lougheed regime and what it represents. In all the talk of rapid growth, economic diversification, massive oil sands development, petrochemical industries, energy corridors and expanding population growth, conservatives rightly sense that the stability of the old order is being undermined. The boom is on and will last so long as world oil prices stay high, stimulating the development of new energy resources and energy related industry. Yet even if the boom lasts, which cannot be guaranteed, it will bring with it its own penalties as well as its rewards. There will be environmental costs to pay, high inflationary and other disruptive effects will be felt in the economy as the result of the huge capital expenditures needed to industrialize the oil sands, and there will be those who will not share in the benefits of growth. Alberta's large native population, living in shocking conditions relative to the standards of white Canadians, is not now sharing the bounty of the natural resource boom and it is unlikely to do so unless it can muster the political power to force some redistribution of wealth. A hint of what may lie ahead for some other Albertans was provided in 1974 when a bitter controversy broke out in the Round Hill area southeast of Edmonton over the plans of Calgary Power to fire a major new power plant with coal stripped from some of the best farmland in the province. Industrial growth requires fuel, and that need sets in motion the familiar chain of events in which we sacrifice renewable resources for the exploitation of nonrenewable raw materials. At some point Albertans, along with other Canadians, are going to have to ask themselves whether the benefits of this kind of growth really outweigh the long-term costs.

If political conflict does develop over the costs of economic growth, the tar sands could well become its focus and emblem. Once a source of frustration because the obstacles to development appeared insurmountable, the sands are now the root of fears that the pace and scale of future development will be unregulated and exceptionally hazardous. Increasingly part of a larger national debate concerning the costs of continued large-scale natural resource development on continental timetables, U.S. influence over Canadian economic and political choices, and the penetration and fragmentation of vulnerable northern hinterlands by the big resource companies, conflict over the right way to develop the immense potential of the tar sands is certain to increase as the consequences of forced growth come home to roost.

Where else but in Canada would an elected cabinet minister responsible for the protection of the natural environment emerge as a public advocate of the strategy of rapid resource development? William Yurko, until recently Alberta's minister of the Environment, articulates in his public utterances the same philosophy of development which underlies the advice of the government's oil consultant, Walter J. Levy. As we have seen, this school of opinion favours swift, conscripted exploitation of the tar sands on political and economic grounds. Alberta, it is advocated, should seize the possibilities opened up by recent shifts in global energy politics to bring the tar sands into immediate production and establish a momentum of development. For unless Canada acts decisively, the major oil companies might by-pass the Alberta bituminous sands in favour of alternative energy resources. Against the risk that the tar sands could thereby become a worthless asset, the cost of rapid development are insignificant. Thus, until the Syncrude crisis erupted in December 1974, the Lougheed government seemed bent on a policy of developing the tar sands at a pace as rapid as the sheer physical limitations would permit. At the same time, it has been committing financial support for environmental research and social programs in the tar sands. One hand of the government is encouraging rapid growth, another is dispensing public monies to clean up the mess created. As the problems involved in the massive industrialization of the tar sands threaten to overwhelm the administration, the government's inclination is clearly to bring order out of chaos by resorting to heavy-handed authoritarian rule.

In a policy address to a 1974 oil sands conference held in Edmonton, Yurko spelled out the Lougheed government's answers to three "fundamental questions." First, should the rapid development of the Alberta oil sands be undertaken? "Today," said Yurko, "the obvious answer is yes. Yet just a few months ago the answer was perhaps not so obvious." A number of factors, foremost among them the new higher prices of crude oil, had tipped the scale in favour of development in a hurry.

Second, under what type of control and organization structure should this development proceed? In the government's view the idea that "this vast resource can or should be developed entirely through public endeavour seems foolish." Such development would have to be done, in the mode of Syncrude, in a "joint partnership between government and industry."

Third, at what rate should the Alberta oil sands be permitted to proceed? Here Yurko was careful to disassociate the government from the half-baked Kahn-Goyer plan of twenty billion-dollar plants in half a decade. The only function that piece of think tank absurdity has had is to make Walter Levy and Peter Lougheed seem reasonable, even cautious by comparison—which they are not. Yurko spoke of "overcoming" environmental constraints on development, adding the pertinent observation that political considerations could make the momentum harder to slow. "It will be difficult politically to hold back development in the face of an idle work force of qualified tradesmen, technicians, engineering and scientific talent." The most "optimistic" rate of development Yurko could foresee had Syncrude's and Shell's surface mining operations coming on stream at the end of the 1970s with a capacity of 100,000 to 125,000 barrels per day per plant. In the early 1980s a plant would begin production every one and a half or two years, accelerating to a plant per year in the late 1980s. Some of these would be mining complexes, others would utilize *in situ* technologies. By the 1990s the rate would continue to be one plant per year, but these would be second generation plants with production capacities of 200,000 to 300,000 b/d. "As a result it is anticipated that the annual production of synthetic crude will reach three million barrels per day from approximately twenty plants by the year 2000."

The first problem with these projections is that the government itself is not in the tar sands business; thus whether development proceeds on a fast scale or not is really beyond its control. If the world price of oil stays at its present high level or rises, then the development of North American alternative fuels could be very fast indeed. Conversely, if the world oil price drops significantly—through the emergence of competition among the oil producers who control eighty-five percent of supply—the incentive to get into high cost substitutes will decline accordingly. The governments of Alberta and Canada can "encourage" rapid development through the use of the traditional resource incentives, and they have certainly greased Syncrude's wheels, but they are refusing to make all the investment decisions themselves; nor can they have much influence on the fundamental question of whether the producing countries maintain their common front—which will require tremendous coordination of production, taxes, etc., over a long period of time. The point is that Mr. Yurko's projections are utterly hypothetical since they are founded on a premise that the highly unpredictable international conditions which brought about the new tar sands play are likely to be permanent.

The second difficulty with Yurko's "optimistic" blueprint is that technically it may well be impossible to implement. The same physical limitations which are turning the great expectations of America's Project Independence into soured dreams will inevitably force Canadians to scale down their ambitions for the oil sands. At present Alberta's Energy Resources Conservation Board, proceeding with its fixed ideas of supply and demand, is staging each surface mining operation at intervals of two years. Others argue that we should build one plant per year if we want to make up for the coming decline in our domestic oil production. In 1972 the National Energy Board forecast that simply to meet Canadian needs, the tar sands would have to be producing 200,000 barrels a day by 1977, 800,000 b/d by 1979, and 1.5 million b/d by 1985. The Energy Resources Conservation Board would like merely 800,000 b/d by 1985. But none of these targets can possibly be met: the only value of such estimates is that they provide us with yet another measure of the incompetence of those who are supposed to be regulating the energy industries in Canada. But who will save us from the regulators? Carleton University geologist Ken North has remarked of the 1972 National Energy Board timetable for the tar sands: "such a program would oversaturate Canada's access to capital markets and our ability to fabricate or purchase steel, cement and electrical components. It would require the services of more engineers, construction crews and machinery than we could possibly manage. No other major engineering undertaking of any kind could be attempted during the duration of this construction, which would bring in its wake terrifying social, economic and environmental consequences." The Energy Board has since scaled down these earlier projections.

Oil sands plants, as Syncrude is discovering, face severe competition for engineers, skilled tradesmen, materials, heavy equipment and construction crews from a host of equally ambitious large-scale North American and European resource projects. Shortages of cement, pipe, steel and heavy oil sands technologies (draglines, hydrotreaters, etc.) and manpower problems—which may require changes in Canada's immigration policies—cannot easily be overcome. There will be corresponding limits in the capacity of the Alberta economy to expand the infrastructure facilities—roads, services, new towns, etc.—necessary for fast development without overheating the provincial economy and jeopardizing every other construction project. In the end, these technical and physical constraints, the sheer inability to build plants faster than they can be built, may be our salvation. Canadians will simply have to adjust to the reality that they cannot build their dreams of self-sufficiency on a tar sand foundation.

And even if it could be done, there would still be overwhelming arguments against the hasty rush to develop the tar sands. The law of supply and demand can be a cruel tyrant. What price are we willing to pay to keep the domestic supply from falling behind the ever rising demand? As we shall argue in subsequent pages, the economic losses to Canada from foreign ownership in the tar sands may run to many billions of dollars. This is intolerable and reason enough to reject the current trend of development. Yet the economic costs do not provide a full measure of what we are risking and in danger of losing in the tar sands. Against the benefits of rapid development—including the drive toward national oil self-sufficiency—must be weighed the heavy ecological and social costs, the exploitation of nature and man, of what is being done in northern Alberta.

Ecological controversies are certain to loom large in the political future of the tar sands. The only major survey of the environmental problems—the so-called "Integ" report of Intercontinental Engineering of Alberta—concluded in 1973 that unless preventative measures are discovered and implemented, "the environmental effects of eventual multi-plant operations over the extent of the Athabasca tar sands could be enormous." A senior provincial official has warned that development could "turn the Fort McMurray area of northeastern Alberta into a disaster region resembling a lunar landscape"; another predicts that the mining technology currently in use will turn the land surface of the sparsely populated muskeg wilderness and boreal forest into a "biologically barren wasteland." Anyone who has toured the devastated lease of Great Canadian Oil Sands can well believe it.

Oil men and government cabinet ministers claim these fears are groundless or exaggerated, but the burden of proof is really on them. The ecological impact of bituminous sands development is unlikely to be gentle, no matter how closely controlled. Here in the dry, rather academic language of the bureaucrat is how the Alberta civil servants summarized the likely effects of future mining and *in situ* projects: "The basic impact on the environment will be partial to total denudation of the surface vegetation, partially disrupted to totally obliterated surface hydrology, extensive changes to the groundwater regime caused by increasing injections and recharge capability modified by a greatly increased permeability rate of the bituminous depleted sands, altered topographical land forms caused by the deposition of spent tailings or the subsidence of depleted sands, massive withdrawals of surface water from streams and rivers causing physical changes to their stream flow characteristics, heated effluent waters resulting in chemical and biological changes to the receiving waters and atmospheric changes such as ice fog during the winter, atmospheric gaseous emissions containing sulphur dioxide and other compounds, all of which will have disruptive effects on the remaining flora and fauna because of the massive ecological changes." In brief, the impact on nature could be disastrous.

All over Canada, continental pressures for immediate frontier resource development are outdistancing the ability of ecologists to plan for the likely consequences of such development. The Science Council of Canada, in a discussion which has striking relevance for the tar sands, remarks that our knowledge and understanding of the fragile northern ecology are most inadequate; crash programs to collect badly needed information, only after development decisions have been made, "will neither relieve the knowledge deficiency nor provide strong foundations for a sound development policy." "Operations and planning for resource exploitation, transportation corridors and centres of population in the North should not proceed ahead of the development of man's understanding of the North or the establishment and use of effective mechanisms to provide protection where necessary." What is required is sustained research support for ecological studies to offset "the increasing pressure to capitalize on short-term profits by immediate exploitation." But research alone is not enough.

In the tar sands there are immediate pressures to develop, yet basic ecological studies are only commencing. In virtually every area of potential concern, rudimentary "baseline" knowledge of the ecological system of the region is almost non-existent. A recent report on the need for study of the effects of development on the Athabasca River remarks, "at present we cannot predict even the simpler hydrologic characteristics of the river, nor have we any real

103

idea of the effluents produced by the oil sands plants.'' Other factors are "poorly known," "uncertain," "open to speculation." What we are doing from an environmental standpoint is walking blind-folded into the industrialization of the tar sands. One industry ecologist remarked in conversation with the author that before commencing development on the scale presently being contemplated, the government should have initiated ecological studies back about 1948 to monitor water flows, climate changes, soil conditions, temperature inversions, etc., on a long-term basis. But such concerns were not taken seriously before the 1970s—certainly they played no role in the decision to allow GCOS to go ahead in the mid 1960s[3]—and now the political and economic pressures to develop are overriding fears about the environment. To be fair, the governments of both Alberta and Canada are in the process of launching major environmental research programs but the essential studies are only now beginning—far too late.

There is a risk that environmental concerns will tend to be resolved by the politics of tar sands development, and this could impede sound planning in several ways. For instance, the inclination of the oil companies will naturally be to pass on as many of the heavy environmental costs to the public sector as it can through cost sharing or by having their costs deducted from their overall tax burden. The nature of the leases and the corporate tendency toward extreme secrecy in matters of proprietary technologies and research could restrict the development of environmental planning on a regional basis—which must, of course, be done across lease boundaries. The large rectangular leases are illogical from an environmental perspective because the boundaries follow survey lines instead of watersheds or natural landmarks. Sound regional planning should involve a careful sequencing of plants, avoiding bunching in a particular location or any other decision likely to compound and multiply the risks. Yet this could involve considerable interference in the development process, possibly deferring or even refusing some applications for development, and there is little sign that the Alberta government is prepared to risk this kind of action.

The earliest tar sands plants will be using the surface mining and hot water extraction technologies to mine the sands and recover the bitumen, and this technique is responsible for a good deal of anxiety among ecologists about the dangers of hasty development. In the recovery process currently being used by Great Canadian Oil Sands and planned, with modifications and larger economies of scale, by the next developers as well, the muskeg must first be drained and removed with all other surface vegetation. Bulldozers, front-end loaders and a fleet of 150-ton trucks tear up and carry away the over-burden, the layer of soils overlying the bituminous sands. At GCOS the exposed oil sands are then mined by two giant bucket wheel excavators, each weighing some 1,800 tons, standing 100 feet high and 210 feet in length, and capable of moving an average 5,000 tons of sands every hour. The tar sands are then transported to the processing plant by rapidly moving conveyor belts. At the processing plant the bitumen is separated from the sands through a scaled up adaptation of the Clark hot water extraction process. Then the bitumen must be upgraded and improved, and this is accomplished through a series of stages involving dilution, centrifuging, coking and the addition of hydrogen. The waste sand and the hot water used in the separation stage, along with small amounts of unextracted bitumen, clay and various chemicals, are diverted into a tailings stream which pumps into a huge lake located behind the plant on the river bank.

This package of technologies combines some of the worst features of intensive strip mining with the threat of serious pollution of the atmosphere and nearby water systems.

In addition, the tar sands present some unique ecological hazards which are going to require unique solutions. The oil sands in the surface mineable area of the Athabasca deposit will see some of the world's biggest strip mines concentrated in a relatively small area. Preparing the sites—as is now happening on Syncrude's lease—involves obliteration of all surface vegetation, diversion of streams and rivers, eviction of all animal life and the ploughing under of any traplines which happen to be in the way. This will cause ''a complete change of land use from wildlife habitat, hunting, trapping, fishing and other wilderness qualities to a biologically barren wasteland,'' in the words of a government environmentalist. The existing natural landmarks will be completely rearranged and major erosion problems created. Much of the topsoil will be used to construct the large dykes for the tailings disposal area. Building the tailings ponds involves further forced changes of the land surface, the transfer of millions of tons of earth and the creation of artificial landforms with steep, unstable slopes exposed to wind and water erosion. Syncrude's plant alone will involve the disturbance of some thirty square miles of land surface.

The stock reply to charges that these truly massive strip mining operations will create a worthless, lunar-like habitat is that the surface will be reclaimed. Government suggests, indeed, that the surface will be more biologically productive after reclamation than before its disturbance. All mining operations must file reclamation plans and post a deposit, and each barrel of synthetic fuel produced will be charged a levy of a couple of cents for a reclamation fund. But most of these expenses will be paid for by the consumer, and there is still no guarantee that the land can be reclaimed. Rehabilitation of the restructured, spent soils in the harsh subarctic climate will involve generations, and no one can in good faith guarantee that it can be done. The reclaimed soils and sands will occupy a far larger volume than in their original state, so the final elevation of a reclaimed area will be sixty to one hundred feet higher than the original contour of the terrain. Unless the reclaimed earth can be stabilized with grasses, shrubs and trees, it will simply erode like a sand dune, and the sparse grasses now growing on GCOS's dykes do not exactly inspire confidence in the prospects for reclamation.

Another severe problem—as with most synthetic fuel projects—is that the existing technology will consume and pollute enormous volumes of fresh water from the Athabasca, only a portion of which can be treated and returned to the river. Disposal of the liquid wastes or tailings left over from the hot water extraction process constitutes the worst single ecological problem in the operation. At GCOS the plant draws in from 6,000 to 9,000 gallons of fresh water from the Athabasca every minute, but it returns a good deal less—the difference being stored in the steadily growing tailings pond. The magnitude of this problem can be grasped from the fact that the tailings ponds being planned for Syncrude, Shell and the other plants will each cover nearly ten square miles of land. The tailings stream is composed of sand, hot water, unextracted oil, fine mineral and clay particles, and some highly toxic chemicals used in extraction. The water is so contaminated that much of it can neither be reused nor returned to the river. Another problem is that the clay particles take a very long time to settle and linger in a state of suspension, thus delaying recycling and reclamation. The result of this could be a truly massive accumulation of oily, polluted waste in large lakes on every developed lease. The GCOS tailings pond sits precariously on the edge of the river, and any serious break in a dyke or seepage underground could cause the ecological ruin of the Athabasca River—a major tributary of the whole Mackenzie system. Whether these oily, heated waste ponds will constitute a hazard to migrating birds is open to speculation. What is certain, however, is that the tailings problem will put pressure on the fresh water supply of the Athabasca: twenty plants would consume up to forty percent of all the river's monthly flow. The planned *in situ* steam injection plants will also consume immense amounts of available water and have an unknown effect on the groundwaters of the region.

The tailings ponds may also have a negative effect on the atmosphere of the McMurray area, and could even alter the climate of the region. The waste water is heated when it is discharged to the pond, and in winter this creates ice fogs in the surrounding area. The Athabasca River valley is characterized by frequent temperature inversions, and under such conditions the atmosphere's capacity to disperse pollution is reduced. Because the plants will be emitting amounts of sulphur dioxide and other compounds, the combined effect of inversions and sulphuric ice fogs could be hazardous to health and to the surrounding natural environment. One government report has alluded to the dangers of creating another Sudbury-style environment in the tar sands and to the possibility of deadly "killer fogs" lying over the river valley. Industry spokesmen tend to dismiss such warnings as alarmist, yet the truth is probably that no one really knows what the total ecological influences of large-scale development will add up to.

The best available evidence that neither the companies nor the government of Alberta have addressed many of these serious problems with sufficient gravity was contained in an Environment Canada critique of Syncrude's "Environmental Impact Assessment." Environment Canada's report, prepared in the summer of 1974, found, according to federal Environment Minister Jeanne Sauvé, that from an examination of the available information (and her staff encountered "great difficulty" in obtaining certain information from Syncrude) the company "has failed to appreciate the real scope of environmental concerns and has also failed to address the question of environmental protection in either a realistic or an adequate manner." Syncrude's documentation "is deficient in detailed information in many areas of environmental concern and we believe that there is a likelihood for major environmental damage." The Ottawa environmentalists noted in their report:

"The Syncrude Environmental Impact Assessment was found wanting in quantitative data relevant to the existing ecosystem components (biological and physical) on Lease 17 and the Athabasca tar sands in general. The functional relationships of ecosystem components lacked quantification and specific aspects of the Syncrude development proposal lacked adequate clarification to effectively predict the ecological consequences of the project. In view of these voids in information, statements presented by the proponent relating to the environmental effects forecast from the development must be considered as conjectural. . . ."

Beyond this rather devastating rejection of Syncrude's entire impact statement the federal report pointed up many specific deficiencies in the consortium's approach to environmental management. "In general, the documents present a concept of pollution management with very little evidence or documentation of reliability," said the federal researchers. This applied to problems of handling huge quantities of waste water, atmospheric emissions, including sulphur, hydrocarbons, nitrogen oxides and hydrogen sulphide. "With the release of large volumes of water vapour, we are concerned with the potential for formation and persistence of widespread fog in the area. This fog, along with sulphur dioxide, could produce a serious human health hazard." The report strongly implied that Syncrude was not employing the best available technologies for reducing atmospheric emissions because these would increase its costs. Likewise, the report noted, Syncrude was making only a token effort at land reclamation: "At no point in the assessment is there any . . . evidence to support the feasibility of reclamation following such a massive physical and chemical alteration to the environment as that proposed by Syncrude." There was criticism of a similar glib approach to the problem of effects on wildlife, particularly migratory birds.

Replying to Mme. Sauvé's letter and to the federal report, Alberta's Environment Minister Yurko commented that "most of the deficiencies identified by your regional task force have also been identified by my staff." If so, these had not been made public, in spite of an Alberta commitment to release all such information. Alberta's policy in regard to the tar sands environment, added Yurko, was that energy supply must have priority: "We know that major information gaps exist in respect to baseline environmental data in the entire area. Nevertheless, in the light of Canada's critical energy balance, it did not and does not appear prudent to delay oil sands development until all needed information is available." The only difficulty with this is that Alberta approved the Syncrude project and laid down the basic environmental guidelines in 1973 at a time when the plant was aimed for the export market: the decision to encourage Syncrude to go, as we shall see, was not taken because of any concern over "Canada's critical energy balance." Yurko's letter is thus a disingenuous and *ex post facto* rationalization; moreover, it is important to note that the tradeoff between supply considerations and environmental dangers—and clearly it is a tradeoff today—has never been thoroughly and openly debated in Alberta, particularly not by Mr. Yurko's government. With both levels of government now directly involved in the project, there is an obvious risk that supply and commercial considerations will be given precedence over the ecology. In February 1975 Ottawa and Alberta announced the commencement of a joint ten-year environmental research program, estimated to cost $40 to $50 million and covering all of the major problem areas. This was the right decision: the big question is whether it is soon enough to prevent major damage.

As with the environment, many of the future social effects of rapid oil sands development are already in an embryonic stage in the Fort McMurray region. The key questions here relate to the social costs and benefits of development, how well development is managed and controlled, and by whom. Impact on the native population of northeastern Alberta can be expected to be especially strong, and here the issue of special compensation arises. Another significant question, to which a depressing answer has already been given, is the extent to which the people of the region will themselves control the future of their own community and the decisions about the coming developments in the tar sands. But let us begin with a brief look at the impact on Canada as a whole.

Recall that the oil companies holding the leases in the tar sands are almost entirely foreign-owned. They are also the same companies which have exclusive concessionary rights in the western sedimentary basin, in the Beaufort Sea, Mackenzie Delta and high Arctic, in Hudson Bay, and in the offshore waters of eastern Canada. Many small "independents" are involved in the hunt for new energy reserves, but the major investment decisions invariably will be taken by a handful of very large corporations which have interests and irons in the fire all around the world. It is these few companies— Exxon, Shell, Gulf, Atlantic-Richfield, Texaco and the others—who will in the final analysis decide whether or not to build the Mackenzie Valley gas pipeline, develop the oil sands or the oil shales, invest in coal reserves or uranium, and so on. Their interests lie in maximizing their overall profitability and in enhancing their long run reserve strength and investment opportunities around the world. Wherever possible, they try to avoid dependence on outside sources of capital, and they also buy from and sell to themselves whenever it is feasible to do so. The large corporations attempt to function as closed economic systems, so they want a free hand to incorporate and license their own technologies, to use their own personnel and market their products as they see fit. Finally, of course, they want to reduce their overall tax burden and have the host country absorb as many of their costs and underwrite as many of their risks as they can get away with.

Given the nature of these companies, rapid exploitation of the tar sands is sure to compound Canada's foreign ownership problem which is already extreme. Each new investment by the big oil companies in a new resource such as the tar sands deepens their overall penetration of the Canadian economy, increasing our burden of foreign ownership and debt, and making a reversal of past and present trends that much more unlikely. The oil companies argue that they are reinvesting today's high profits in the development of tomorrow's energy reserves, and to a degree they are. But the problem is that they are also thereby reducing our future economic possibilities. It has been demonstrated that because of the high degree of foreign ownership in our branch plant oil industry, a net loss accrued to Canadians every time the price of oil rose between the early 1960s and 1973. What is happening in the tar sands and in other reserves which will one day replace that oil is that the same kind of heavy economic burden is steadily being fastened onto tomorrow's resources. But there may be a difference of degree. If development proceeds at the fast pace and on the very large scale now being planned for the tar sands and northern natural gas, etc., the loss of economic wealth because of foreign ownership will be on a far grander scale than anything we have yet seen. Moreover, government's ability to cut these losses will be weakened to the extent that the decisions to accelerate development are left in the hands of the private sector. Syncrude's success in extracting concessions from the Albertan and Canadian governments derives from its monopoly veto power, from its capacity to put a spoke in the wheel of development merely by saying "no go."

Very large investments of capital, materials and manpower in the tar sands must also be balanced against Canada's other pressing economic and social priorities. If the nation is to have anything left over for such needs as housing programs, education, day care, an attack on the disgraceful living conditions in which Canada has left its large native population, and the development of job producing industries, then we shall certainly have to cut back excessive investments in big energy projects such as the oil sands ($2 billion plus per plant, with up to twenty plants projected over twenty-five years), the Mackenzie Valley pipeline ($7 billion plus), the Polar Gas pipeline from the high Arctic (another $7 billion plus) and the James Bay hydroelectric development (up to $12 billion). The shopping list for these large northern resource projects is long and carries a staggering price tag. It is true that tar sands development will create many more permanent jobs than a pipeline, but it still costs well above a million dollars to create one permanent job in a plant like Syncrude, compared to a figure between $50,000 and $70,000 in manufacturing. Canada simply cannot afford more over-development of her resource economy: such an economic strategy is short-sighted, for it will increase our burden of foreign ownership, deepen our political dependence on the United States, and further weaken our already stagnant manufacturing sector. Trade surpluses built almost entirely from the export of resources in raw and semifinished form are notoriously susceptible to downturns in the international economy. In highly manufactured goods—where innovation, efficiency and technological know-how count—Canada has a huge deficit which expanded from $3.5 billion in 1971 to $6 billion in 1973. In 1974, Canada's deficit in end products trade increased to a record $9.5 billion. It is here, and in policies aimed at redistributing wealth, where Canadians should make their real investments in the future. Excessive resource development is a recipe for more dependence and social inequality.

Understandably, there is a good deal of support within Alberta and the rest of western Canada for the creation of new industry. The fetish for economic growth in Alberta is partly due to the undeniable reality of depleting oil and gas reserves: a widespread and very appropriate demand has arisen that before the wells run dry the province must make arrangements for the future. Peter Lougheed has very shrewdly linked his political strategy to this demand and to the related historic grievances of westerners against central Canada. International events entirely beyond his control have played into his hands to an extent no one could have foreseen when he and his "NOW" Conservative party swept into power over the spent corpse of Social Credit in 1971. The problem is that "more industry" and "better value for our resources" have simply become slogans—indispensable for safe after dinner addresses to the Chamber of Commerce but not exactly a blueprint for the good society. Why develop industries? What kinds of industry? How much? How fast? At what cost? For whose benefit and at whose expense? These crucial questions remain largely unasked. Western alienation has become another motherhood issue. It is ironic and depressing that many of today's westerners who are the loudest in their denunciations of eastern exploitation are among the least exploited, most privileged people in the world. Western resentment has many legitimate roots. But it is fast becoming the illegitimate blunt sword of those who would play off one part of Canada against the others, keeping old wounds open and old animosities alive, diverting attention from real injustices to imaginary enemies. It is an old and vicious game and far too many Canadians seem prepared to play it indefinitely.

If eastern Canada did not exist, what kind of society would westerners create? An answer to that question may lie in the ways Albertans distribute the costs and benefits of the development of the tar sands. Will the northern parts of the province be treated simply as a resource rich colony? How much will the province's forgotten native people receive from oil sands extraction?

, In their perennial obsession with east-west confrontations, most Canadians have conveniently overlooked the fact that the real lines of social, economic and political exploitation within the country increasingly follow a north-south pattern. Power in Canada resides in the south—the closer to the 49th parallel the stronger it is—power resides in the cities, in that part of the population which is urbanized, consumption oriented and North Americanized, the population which dominates the cultural, economic and political life of our cities, our provinces and the country. By contrast, the North—including the northern peripheries of almost every province—is powerless, depopulated and poor. Canadians view their North like the British used to regard their empire—from a great distance—and, adopting a similar paternalistic colonialism, Canada's southerners have normally been content merely to administer the North by remote control. The energy crisis, however, has forced Canadians to re-examine their collective consciences, and we have now decided that what the North actually needs is "development." We have reformed, we can grasp our moral obligation to upgrade our northern ghettos, to give them the CBC and a steady stream of anthropologists in return for some of their underdeveloped resources. We are all in agreement that it would be criminal of us to continue to ignore the North. One day northerners will even be ready for self-government; until that day we can at least give them growth.

If the Lougheed government's vision of twenty oil sands plants is ever realized, northeastern Alberta is going to get a very large dose of administered growth. Already Fort McMurray is straining at the seams, and some projections have the town expanding from its present size of about 15,000 to some 70,000 people in the next two decades. That expansion could cost the equivalent of half of one tar sands plant—perhaps a billion dollars—and there has been serious talk of a completely new town, another McMurray, being implanted north of the present townsite to service the Shell, Petrofina and Home Oil projects. Syncrude's development alone will require Mc-Murray to build more than 2,000 new housing units, double its hospital capacity, double its commercial property, treble its school classrooms, build new recreational facilities, and develop whole new suburbs. These kinds of demands are placing fantastic pressures on the townspeople. There is a permanent housing crisis, reflected in the exorbitant rents workers are forced to pay and in the excessive values of privately owned property. Edmonton and local developers have exploited the scarcity crisis for all it is worth; in a typical instance, one developer made a 400 percent profit by picking up a piece of land and selling it to the local school board for a new high school site just eight months later. Space in the trailer camps is at such a premium that some families live in tents on the edge of town in summer, then pack up when it gets too cold to stay on. The many trailer parks are overcrowded, lacking in playground facilities, with muddy or dusty streets, depending on the weather, and still the stalls rent for more than in Edmonton. One individual who brought a trailer from Edmonton was charged $500 by a local firm for a place to park it, and an additional $300 for utility hookups.

McMurray's future has become something of political football, the town plan has been scrapped, revived and revised, and there have been charges that cabinet ministers in the Lougheed government have been indulging in rampant patronage in awarding consulting contracts. The working people of McMurray have never had much to say about all of this: since the mid 1960s the town has been under the direct control of the province (under the New Towns Act), and now it is being put under the one man rule of a regional commissioner who will be responsible only to the cabinet and who will have the authority to by-pass or override much existing legislation. This resort to authoritarian wartime solutions and strong press reaction to the policy clearly embarrassed the Conservative government and it is perhaps the strongest indication yet that the pressures for tar sands development are already getting out of control. During debate on the bill enacting the commissioner concept, one Tory cabinet minister was moved to remark that, "if we find that we've put an Adolf Hitler into Fort McMurray, we are quite capable of taking him out again." Another minister recalled being on hand for the start up at the GCOS plant back in the 1960s; it was like "the beaches at Normandy on D-day. That was only a $300 million project. How on earth can the normal infrastructure of local government and the normal channels of the civil service be expected to absorb the impact of a second plant three times as big, and another, and another, and another, every two or three years?'' Another Conservative was ''proud to be a member who is going to support this bill, because if we do something wrong now, at least we are doing it.'' Retorted an opposition politician, ''I was more enthused about the bill before I heard the ministers speak to it.'' The people of McMurray, Fort McKay, Anzac and the rest of northeastern Alberta have reason to be even less enthused.

Nor have Alberta's native people, some ninety percent of whom are undereducated, unemployed and forced to live in abominable housing, little reason to be excited about the prospects of rapid tar sands development. Alberta's supposed affluence is not to be found among the shabby houses and mud streets of Fort McKay, just a few miles north of the Syncrude site. Alberta's natives have been by-passed by the province's oil and gas boom and they have scant hope of deriving much benefit from Syncrude and succeeding projects. The government has been making an effort to see that native manpower is utilized in the oil sands. But native housing and dis-crimination in the McMurray area are bad or worse than any in the province. Far from benefiting to any significant extent from rapid oil sands development, Alberta's large Indian and Métis population could be victimized by such a policy—unless they fight back. The Indian Association of Alberta is said to favour a hardline policy toward a major land claim in the tar sands area, covered by Treaty No. 8, and the subject of similar action in the Northwest Territories. Association president Harold Cardinal has vowed to launch a claim and to stall development with legal action if necessary to win an ade-quate share of the resources for native people. Cardinal points out that while the province's native resources are completely wasting, the oil companies are talking of importing labour to overcome the manpower shortage.

Whether the tar sands benefit Albertans and Canadians—par-ticularly those who need it most—or the foreign-owned energy com-panies, will be decided by political power, not by pious corporate codes and rhetorical speeches. A great deal of social, as well as eco-nomic value can be squeezed out of the tar sands, but at present Ca-nadians are probably losing more than they are gaining by proceeding with development.

8   The Politics of Syncrude
I: The Rules of the Game

'Of all the commodities moving in international trade, oil undoubtedly is the supremely political one. Because most oil moves internationally, because the trade is of enormous monetary value, because huge profits are to be made, because it is a vital necessity to most oil-importing countries, because it is of crucial importance to the economies of the oil-exporting countries and the balance of payments of the developed countries—for all these reasons, the ordinary day-to-day flows of international oil trade are the resultants of enormous and conflicting political pressures among the companies, governments, and international organizations.''

Michael Tanzer
The Political Economy of International
Oil and the Underdeveloped Countries.

The attitude of Canadians to the political activities of international corporations operating in their country can be likened to the Victorian approach to sex. From time to time we hear vague rumours that something indecent may be going on, but we prefer not to discuss the matter, certainly not in public, and it is generally understood that those who pry into such matters are behaving in the worst possible taste. A prudish Sunday school mentality about the political facts of life pervades the entire society, beginning in the schools and universities where the subject of corporate power is tactfully kept off the curricula, to the business pages of any newspaper where the latest ruminations of any branch plant junior vice-president on the perils of Canadian nationalism may be read as the last impartial word on the topic. Eric Kierans, the outspoken ex-cabinet minister who now teaches economics at McGill University, has charged his fellow academics with displaying a "hear-no-evil, see-no-evil" attitude toward the multinational corporation. But the ivory towers only reflect a more general complacency; for the vast majority of Canadians, business is only the business of business.

The country as a whole has paid a high price for its willingness to tolerate, indeed applaud, the quiet dealings of its political leaders with the good corporate citizens of General Motors, Exxon, ITT and scores of others. But the price cannot be determined by an accounting manual, nor can it be retrieved through higher taxes or royalties, since the assets our politicians have alienated have so often been of an intangible nature. How, for example, can we assess the market value of a government's promise to rewrite its laws if necessary to ensure "labour stability" for a multinational corporation? Or how do we measure in dollars and cents a commitment by elected politicians to involve foreign companies in the process of drafting future legislation relating to the environment? What has been slipped across, or under, the bargaining table in such cases is not lost "economic rent," but the less tangible and far more valuable asset of sovereignty—the right and power of a people to make their own choices, to design their own future. Sovereignty: it is an antiquated, almost reactionary sounding concept, a sixteenth-century idea which fits awkwardly into our supranational twentieth century. Yet, until organized labour can achieve the kind of cross-boundary cooperation which can countervail and neutralize the homogenizing impact of the great capitalist companies, the espousal and defence of national sovereignty will retain legitimate roles in contemporary politics. It would be far less legitimate if a truly international entity were emerging in place of the nation-state; but, alas, what presently threatens to undermine nations is not the brotherhood of man. Propagandists of the multinational corporation preach a rhetoric of togetherness, interdependence and cooperation, but the name of the game is still profit making, control, dominance—in short, power politics. Far from overcoming inequality between the rich and poor countries or within Third World countries, the multinationals build on the best and the strong, enriching only a few, distributing growth, income and the benefits of development unevenly. Poverty is created alongside wealth, perpetuating inequality. Moreover, the multinationals are directly involved in the struggle for political power. It may be argued that the international corporation's primary interest remains that of profit maximization: agreed, but in its overall calculations of rates of return the typical large corporation takes into account a host of factors which lie in the realm of politics. Multinational corporations are political institutions *par excellence,* and none more so than the giant companies engaged in the worldwide quest for that most political of commodities, oil. Canada's tar sands have become a focal point of this political struggle.

To uncover the politics of Syncrude in its dealings with the governments of Canada and Alberta we must first pierce the veil of secrecy which encloses the large corporations, shielding them from public scrutiny. Corporate spokesmen react with pained outrage at the suggestion that their activities are in any sense a proper subject for broad public inquiry and investigation, and far too often our politicians, editors and reporters appear to concur. Syncrude, ever concerned about its shiny corporate image, has published an account of its earlier negotiations with the Lougheed government in one of its slick brochures, "1973: Year of Decision," but in spite of the breathless prose ("Meeting followed meeting. Royalty formulae were devised, then torn up again . . ."), little of substance is revealed. The government has been even more reticent, perhaps with good reason. Still, it is possible to reconstruct an incomplete but generally consistent account of the corporate-government negotiations as they unfolded between 1972 and 1975, together with the details of a few of the deals struck along the way. A few pieces of the jigsaw puzzle are missing, yet the essential outlines are all too clear. What we perceive when we examine the politics of Syncrude is more than just another resource sell-out—though Syncrude is unquestionably that. We can also see how foreign corporations operate inside the Canadian political system while holding that system in scarcely-concealed contempt. We see a pattern of corporate bargaining strategies and tactics emerge from the struggle of the private sector to extract maximum concessions from government, and we see too how governments reduce their own bargaining power by foregoing the option of public ownership and development. We discern a highly unequal bargaining process, a built-in inequality which guarantees the kind of result Canadians have come to expect when their politicians sit down at the same table with the multinational corporations. And through all of this, we begin to make out a frightening answer to the unspoken question of what all of this implies about Canadian politics and society.

We encounter three central actors in our analysis of the politics of Syncrude. The actual negotiations took place at the bilateral level and in separate stages, but let us imagine, for the sake of clarity, a triangular negotiating table, along whose sides sit the bargainers of

Syncrude Canada Ltd., Peter Lougheed's Conservative government of Alberta, and the then minority Liberal government headed by Pierre Trudeau.

Syncrude, the corporate actor, was in its original form, the hybrid offspring of four American-controlled petroleum companies. In some of big oil's joint ventures a group of companies will band together for common purpose, designating one company to act as operator for all: such a venture is the Athabasca Oil Sands Project consortium planning the fourth tar sands plant, in which Petrofina holds controlling interest and is taking the operating role. But in the Syncrude venture the participating companies put together a separate corporation and run it jointly. Several of the famous joint ventures in the Middle East, including the world's largest producing company, the Arabian-American Company (ARAMCO) of Saudi Arabia, are of this type. The advantage of the latter approach is that the participating companies are not required to delegate much decision making authority to a competitor and they can exercise complete control over the affairs of the joint company. What this can mean for a host country was vividly demonstrated by the very first of the joint ventures, the old Iraq Petroleum Company, which was created by several U.S. and English companies and operated, at great cost to the nation of Iraq, as a non-profit company for years. In order to avoid taxation in both England and the United States the owners chose to run the company at a loss, transferring profits abroad to other branches of their global organizations. The history of oil is littered with such cases, and Syncrude could well end up in similar circumstances, running up heavy losses initially and never showing impressive returns. As we shall see, Peter Lougheed's profit sharing scheme creates an irresistible incentive for the accounting departments of the owners to keep their costs up and profits down, and there are many tested techniques open for them to utilize.

As its president, Frank Spragins, freely admitted to the author, Syncrude is merely a cost company, operating without independent assets or revenues on "a glorified expense account" for the partners—originally Imperial Oil (thirty percent interest), Gulf Oil Canada Ltd. (ten percent), Atlantic Richfield Canada Ltd. (thirty percent), and Canada-Cities Service Ltd. (thirty percent). Syncrude itself owns nothing; at the end of each month it totals its bills and sends them to the participants, and when synthetic oil begins to flow from its complex on lease No. 17 it will be parcelled out in the same way, each participant taking the percentage it holds in Syncrude. The important point to bear in mind is this: Syncrude is not an independent, profit making corporation. It is merely a cardboard, store-front company, the creature of its owners' interests, enjoying the appearance but not the substance of power. What lies behind it, however, is something else again.

As originally conceived Syncrude was a typical oil industry joint venture, bringing together two senior partners and two junior members of the international cartel. Of the original four, Imperial and Gulf are Canadian affiliates of Exxon Corporation of New York and Gulf Oil Corporation of Pittsburgh respectively, two of the famed seven sisters of world petroleum, while Atlantic Richfield and Canada-Cities Service represent the Canadian holdings of two smaller, aggressive U.S. companies with an intriguing history of involvement in each other's affairs. Back in 1962—four years after the old Richfield Oil Corporation acquired a huge 25c per acre block of tar sands leases along with Cities Service and Imperial—the U.S. Justice Department filed anti-trust suits against Richfield, Cities Service and Sinclair Oil Corporation. The actions were based on the fact that Cities Service and Sinclair had held controlling interest in Richfield since the 1930s, that there were interlocking directorships across the three boards, and that the companies refrained from direct competition with each other in the United States. Cities Service held a thirty percent interest in Richfield when the two joined what is today the Syncrude group, but the anti-trust action subsequently ended this happy relationship, forcing Richfield into a shotgun marriage with the Atlantic Refining Company in 1965 and creating (in the mode of most American anti-trust actions) a brand new energy conglomerate—Atlantic Richfield or ARCO. Richfield, incidentally, had been the inspiration behind an idea seriously considered in the late 1950s to use nuclear explosions as an *in situ* process for extracting oil from the deeply buried sands. An ambitious, fast-rising company, ARCO struck gold in 1968 when it located vast oil reserves on Alaska's north slope, and since that date the company has undertaken a twenty percent commitment toward the building of the Prudhoe Bay-Valdez Alyeska oil pipeline. Alyeska, long delayed because of environmental resistance, has proved to be a costly burden for ARCO, which has also been a participant in oil shale as well as tar sands development. Financially overextended, ARCO was later to emerge as the weak link in Syncrude, but in 1973, as serious discussions began, the four partners were tightly united. And a powerful group they were too: the four U.S. parent companies held combined 1973 assets of more than $40 billion! This enormous financial clout, their control of key technology, combined with their worldwide reserves of oil and other energy forms, gave the Syncrude partners a truly formidable hand in their poker game with the governments of Alberta and Canada.

Syncrude's decisions are made collectively, with the joint executive and management committees, made up of Syncrude's staff plus executives of the participants, reporting to the owning interests. Power resides in the participating owners and relatively little decision making authority or legal power is delegated to Syncrude's operations staff. All the owners take an interest in Syncrude's affairs, but within the consortium the senior partner is clearly Exxon's Canadian affiliate, Imperial Oil. Imperial is Canada's largest energy company, the largest lease holder in the oil sands and active in *in situ* development at the Cold Lake deposit. Several of Syncrude's top personnel, including Frank Spragins, are ex-Imperial/Exxon employees. Imperial Oil, of course, is fully integrated, controls another fifty Canadian companies, produces crude oil, natural gas and natural gas by-products, owns tankers, pipelines, several refineries and markets everywhere in Canada. The company, in line with the deepening vertical and horizontal integration of all the big oil firms, is one of Canada's biggest chemical manufacturers and is now engaged in the purchase of uranium and coal properties in a bid to monopolize tomorrow's fuels. Imperial's corporate strategy appears to envisage special concentration on the tar sands, the Mackenzie Delta and the Beaufort Sea in the Canadian Arctic. Imperial's predominance in the tar sands reflects its seniority within the branch plant Canadian oil industry as a whole, a seniority which also gives its top executive officers the unwritten status of the industry's leading political strategists.

Interestingly, Imperial is not the only Exxon affiliate with heavy commitments in the Syncrude consortium. Back in 1972 Syncrude initiated some dozen or more purchasing agreements with Esso Research and Engineering, a 100 percent American-owned affiliate of Exxon, adopting for the upgrading of bitumen the patented hydro-treating and fluid coking technology developed by Esso Research. Esso will thus be heavily involved in the design and construction of much of the last "refining" stages in Syncrude's plant operation. Such are the mysteries of vertical integration that the licensing of Syncrude's huge fluid cokers and other expensive refining technology and expertise from Esso Research and Engineering will transform some of Syncrude's costs into revenues within Exxon's corporate system: costs absorbed in Alberta will reappear as profits in the United States. Syncrude does not talk much about the nature or cost of its dozen odd purchasing agreements with Esso Research and Engineering; such information, callers are informed, is "proprietary." It is also "controversial."

How much autonomy of decision making power does an affiliate such as Imperial Oil enjoy? The question is important since Imperial's executives have always taken pains to argue that Exxon/Jersey exercises little or no direct control over their major investment decisions, pricing policies, marketing strategies, and so forth. Imperial claims, indeed, that it has the effective independence of a national oil company, and that its parent functions as a passive holding company, interfering very little in Imperial's day to day operations. Given the tightly knit nature of the international oil industry, this has always seemed an unlikely proposition; but until the Nova Scotia Power Corporation launched a $100 million lawsuit against Imperial over a fuel oil supply agreement, the evidence was unavailable to disprove it once and for all. The court case between the NSPC and Imperial, heard before a justice of the Nova Scotia Supreme Court in the spring of 1975, ended any remaining doubts. While the case centred primarily on the issue of whether Imperial was entitled to pass on to its customer, the power board, the burden of tax increases imposed in Venezuela, the evidence revealed much about the way Exxon directly involves itself in Imperial's pricing and crude purchasing policies. "The judges found," Oilweek reported, "that Imperial does not have independence of action in deciding where its crude supplies come from or an ability to bargain directly with the supplier. This function is carried out by the planning and analysis group of Exxon International in New York." Imperial has no representation among this small group, thus the company cannot control its own source of crude supply. "The ultimate decisions rested with Exxon Corporation," concluded Justice L. S. Hart. "Whether the crude oil that eventually ended up at the Dartmouth refinery would have come from Venezuela, Africa or the Mideast was something that could not have been predicted with accuracy by either of the parties at the time this contract was executed." The case also documented how Imperial used a Bermuda store-front subsidiary to cream more than $35 million in profits from paper transactions involved in moving oil between itself and other Exxon affiliates: Imperial paid neither Bermuda nor Canadian income taxes on this money, transferred to Toronto as offshore dividends.

Imperial has always been the pacesetter in the pricing of Canadian crude and refined products, but the Nova Scotia lawsuit demonstrated that this function too is determined by Exxon. Among the

40,000 documents assembled for the case was a Telex message sent from Exxon International in June of 1973 instructing the company's worldwide affiliates to increase prices and advising them of the desirability of making further "market adjustments." Before Ottawa stepped in to freeze prices in September 1973, crude prices had increased by 95¢ a barrel, or over thirty percent, in less than a year; and each price rise was initiated by Imperial, no doubt as part of Exxon's global policy.

This evidence effectively lays to rest the lingering myth of Imperial's autonomy; neither it, nor any other foreign-controlled oil company operating in Canada, enjoys independence of action, particularly in areas such as the tar sands where billions of dollars are at stake. The companies behind Syncrude are American—none other. Syncrude is about as Canadian as the stars and stripes.

About a year before its first round of negotiations with Syncrude reached the critical stage, the Lougheed cabinet had been briefed by its leading civil service advisors about what to expect from the multinationals holding tar sand leases. The civil servants had warned the politicians that the multinational companies would oppose any attempt by government to maximize Canadian benefits from development of the great resource. For such a strategy would limit both their profits and their future control. Their interests "lie in the rate of return on their investment within their entire corporate structure spread across many parts of the world." The oil companies had been sitting on their leases, waiting for the right moment to develop the tar sands and they consequently monopolized the lease potential information, the technology and capital. They would not be interested in questions such as Canadian economic sovereignty and would wish to purchase much of the construction and operating equipment outside of Canada. "Many of the senior staff positions both in design, construction and operations are filled by non-Canadians," and much of the synthetic crude would be shipped out of Alberta in an unprocessed form. Moreover, the multinationals' interest in maximizing their rate of return on investment would lead them "to externalize as many of the costs arising from the projects as can legitimately be done. Since the environmental costs of the development are extemely high and since the current technology and economics of extraction are still in their operational infancy, the tendency will be for the corporate structure to externalize these costs for society to absorb." And not only, we may add, the environmental costs.

Recognizing the likelihood of resistance to their pro-Canadian policy for the tar sands, the bureaucrats had proposed a strict and activist campaign of corporate surveillance and control. In 1972 the Lougheed cabinet had set down some preliminary guidelines for the Syncrude project, calling for the use, "wherever practical and reasonable," of Alberta engineers, constuction firms, equipment and labour, and, "insofar as it is reasonable to do so," processing of tar sands by-products within the province, as well as public participation by citizens of Alberta. But the civil servants politely dismissed these as vague and meaningless. What they advocated was an overall development policy designed to maximize the use of Canadian capital, labour, technology and expertise. The obvious reason to push Canadian ownership was to retain the economic surpluses in Canada, increasing the pool of capital for more Canadian ownership of future projects. But the argument for stressing the use of Canadian scientists, engineers and other skilled technologists and Canadian technology in the design, building and operating of the plants went beyond the direct benefits of employment income. "The development of technological expertise, once started, is seen as a self-fulfilling or self-feeding process. It stimulates new technologies and is seen as the basis upon which the viability and vitality of economy and society depend." If Canadians were in control, the civil servants went on, they could utilize this knowledge to further the economic and social development of the country. "Industrial and economic development are ultimately dependent upon technological know-how. Canadians and especially Albertans can utilize the development of this unique technology as the lever to further the development of new technologies and ultimately the development of the society." It was the longer view that the Alberta civil servants primarily had in mind. "Thus, the intimate involvement of Albertans and Canadians in the development of the tar sands is seen as a means by which to develop Canadian technology, the Canadian economy and ultimately the Canadian society. It is viewed primarily as a development lever," they told the cabinet.

To implement their far-sighted "development lever" concept, the civil servants wanted the government to scrutinize the existing Great Canadian Oil Sands plant and the proposed Syncrude project to monitor "the Canadian content deficiency" of the basic technology, materials, parts and equipment, and to determine whether Canadian alternatives were available. They also wanted to explore with the federal government the feasibility of restricting the use of foreign capital, personnel, materials and technology in the tar sands. They wanted to force the corporations to file research with the Alberta Research Council, reversing the confidential relationship between the council and the oil companies. In short, they were not proposing a public relations campaign to give foreign control some made in Canada window dressing. They were arguing that the government should set the terms and conditions, the rules of the game, for tar sands development. What they were sponsoring was a tough policy bound to put the government on a collision course with the multinational corporation.

The policy was rejected. It was too risky. It challenged many of the basic ideological assumptions of the governing Conservative party; moreover, the cabinet was in no political position to take on the oil companies. Lougheed's government has on many occasions snubbed its civil service, which it inherited from the long rule of Social Credit, by hiring outside consultants such as Walter Levy's firm to advise it on all sorts of issues. Several firms have been engaged in various aspects of tar sands research and planning including environment, infrastructure, planning regional growth, communications, new townsites, etc. It is shrewd political strategy, for such "expertise," hired on a contract basis is almost certain to tell a government what it wants to hear. By contrast, the frustrated nationalism of the civil servants must have had a painful effect on the ears of the oil lawyers and businessmen of Lougheed's cabinet, men long accustomed to working with and for the Calgary oil industry.

Lougheed's own power base is in Calgary; he is part of its milieu, and shares its style of operation and frame of reference. The men who spoke for Syncrude were, in fact, Lougheed's type of people, his natural allies, far more so than the central Canadian, federally-oriented government so heartily disliked by Calgary's elite. The members of the Petroleum Club had little affection for those who would "hoard resources" or deny the U.S. access to Alberta's energy. Nationalists were worse than socialists; did they not see that Alberta had the chance to become a petro-chemical hub of North America—of the world? Another Texas! Calgary would become the new Dallas, instead of its supplier of computer data and dividends.

What Peter Lougheed articulates so well is the politics of resentment, the frustrated aspirations of a second-tier elite for so long dismissed as boorish cowboys, as yahoos with dung on their boots, by the smug, ruling Anglo-French establishment of Ontario and Quebec. There is some fringe separatism in Calgary, but what the rich and powerful of Alberta really want is a new political balance within Canada. Peter Lougheed is attractive because he preaches growth, rapid change, access to world markets and above all, recognition for the West. The cautious, sceptical nationalism of the civil servants was not likely to impress the premier, his colleagues or his backers, particularly since it pressed for surveillance and control of an industry which Lougheed and many others genuinely believe is responsible for Alberta's present strength and prosperity. "Without a strong and vital oil and gas industry in Alberta, let's face it, we'd be a have-not province," Lougheed remarked during his televised Syncrude announcement. "Our position would be comparable to that of Saskatchewan or Manitoba." That, of course, presumes that the oil industry itself gave Alberta its oil, a confusion the oil companies do their best to propagate; but not even Exxon has been around that long.

What Lougheed and his cabinet had in mind for the tar sands was clearly signalled in November of 1972, a few months following the call of the civil servants for a "development lever" strategy, when Syncrude and its general building contractor, Canadian Bechtel Limited, a subsidiary of the giant Bechtel Corporation of San Francisco, made a slick presentation to the Lougheed cabinet on the economics of the project and the use of Canadian engineers. "Why isn't it feasible to turn the job of managing contractor over to a consortium of Alberta engineering firms?" asked Bechtel and Syncrude. Because such companies lack size, experience in handling big projects and the necessary staff. "Alberta engineers generally lack experience in construction labour relations, planning and logistics for very large projects, since their activities have generally been confined to the engineering phase of projects; with rare exceptions, Alberta firms are not willing to provide large financial guarantees related to design or performance; few Alberta firms have broad experience with debugging, start-up or the environmental engineering of process-type facilities." Moreover, let's not forget, "the major companies in the English-speaking world have had offices located in San Francisco, Los Angeles, Houston, New York, and London, England." If a big airplane were designed by these local outfits, the two American-owned companies added, the airplane's left wing and tail assembly would not match up with those on the right. Or, as another Bechtel official in charge of construction of the James Bay hydroelectric power project summed it up shortly before his workers replied by burning down the site, "Canadians know fuck all."

That was right to the point but a bit indiscreet, and such indelicacies are seldom uttered to governments, at least not by companies with offices in San Francisco, Los Angeles, Houston, New York, and London, England. Understandably, Alberta's engineers have been less impressed and regard Bechtel's control of Syncrude's construction as an insult to their professional abilities; some of them have also had some unkind things to say about Bechtel's own work. But Bechtel is the oil cartel's construction company, still family-owned and the largest engineering and construction firm in the world, and the monopoly position of the oil companies over the leases virtually guarantees it an inside track to the tar sands. Bechtel has worked with the majors and other large resource extraction companies all over the globe, from copper mines in Indonesia and iron mines in Brazil to Cominco's "Black Angel" lead and zinc mine in remote northern Greenland. It has reaped tremendous rewards from its participation in Canadian resource exploitation: contractor for GCOS and Syncrude, the James Bay project, the Churchill Falls complex in Labrador, the extension of the Interprovincial Pipeline to Montreal, refineries, pulp and paper mills, and on and on. Bechtel will soon be building gas and oil pipelines from the Canadian Arctic to U.S. markets. Representatives of the company first appeared in the tar sands in the late 1940s and Bechtel has taken a special interest in the resource ever since.

The firm's chief officer is Stephen Bechtel, Jr., a director of General Motors, Hanna Mining and Southern Pacific Company, and obviously a man of standing among America's corporate-financial elite. The corporation maintains offices in Canada and—in spite of its less than subtle approach to public relations—hires a few Canadian engineers as sub-contractors. But most of the basic design for Syncrude's extraction plant is being done at head office in San Francisco. Syncrude's will be the prototype plant for all future tar sands mining operations, thus Bechtel will almost certainly build plants for the other developers. There is no indication that the provincial government has the slightest intention of restricting Bechtel's monopoly over the important and lucrative construction and engineering side of the tar sands business.

The reluctance of the Lougheed cabinet to challenge the arrogance of the Syncrude-Bechtel *fait accompli* of November 1972, on the decisive issue of Canadian technology and engineering, a key theme of the strategy of the civil servants for repatriating the tar sands, obviously made a mockery of the government's vague Canadian "guidelines." "We have truly taken a great tumble from the pioneering work of Dr. Karl Clark of the Research Council of Alberta," lamented the Edmonton *Journal* reporter who leaked the story. More significantly, the incident signalled the cabinet's willingness to allow the corporate players to keep the initiative and define the rules of the game. The government had flashed the message that it felt itself to be in no position to haggle about the fate of Canadian engineers: the essential thing was to keep Syncrude in the tar sands picture. The consortium's application was then the only one under active consideration; the other lease holders were biding their time, waiting to see what kind of terms the government was offering. Because Syncrude's agreement with Peter Lougheed would establish key precedents in areas such as royalties and public participation the consortium was, in effect, bargaining for the entire oil industry. The government was not taking seriously any suggestion that it develop the tar sands under public ownership, and the majors had effectively barred new entrants by buying up all the prime leases. And so, from where the government sat, the entire future of the tar sands appeared to hinge on the willingness of Syncrude to proceed with the development of lease No. 17. To Lougheed and his colleagues the fate of the tar sands and the fate of Syncrude's application had become inseparable. It was fatal misconception, and one which the owning companies in the consortium exploited at every turn, laying down large and not so large conditions, requesting concession after concession, passing on risks and costs to the public sector, and always daring an unimposing government to test its ability to develop the tar sands without them. Nearly a year before the last hand in the poker game was played out, Syncrude had already convincingly called Peter Lougheed's bluff.

The outcome was predictable. The tar sands negotiations of 1973 involved a large package of issues, essentially Syncrude's exorbitant price for its "go" decision. Some of the items on the shopping list were vital, others far less consequential, but taken as a package—as Syncrude took them—they added up to an extremely lucrative deal, "a sweetheart deal," the federal minister of Energy, Mines and Resources rightly called it. It included a guaranteed rate of return on investment, royalty-free holidays, commitments to provide strike-free labour, promised provincial support in Syncrude's negotiations with Ottawa, expensive publicly financed "roads to resources" infrastructure, and more. Then toss in additional commitments from a generous federal government to permit Syncrude major reductions in its taxable income and the probable privilege of charging international prices. Some of the deals were later made public, others concealed or dressed up in camouflage. Together, they mortgaged the future of one of Canada's most important resources. Enough was eventually given away to win Syncrude's "go" approval and to send the other leaseholders scrambling to get in line. Their eagerness had been triggered by a sell-out, and that indeed was the larger significance of the Syncrude deal.

The vital issues in the Syncrude negotiations with both provincial and federal governments all revolved around the consortium's rate of return on its investment. Royalties were the crux of the matter in 1973. Syncrude argued that the traditional gross production royalties of resource industries, in which a fixed percentage of the value of the produced commodity (for example, twenty-five percent of the wellhead value of a barrel of crude oil) is set aside for the public treasury, should not apply in its case. Syncrude pointed to Great Canadian Oil Sands, arguing that it had been crippled by the necessity of paying royalties during the years of no earnings (in fact, GCOS royalties have been very low and "forgiven" for several years of the plant's life). Syncrude therefore rejected the Alberta government's initial royalty proposal, made in April 1973, for a royalty based on the theoretical value of the raw bitumen extracted from the tar sands. Syncrude wanted a royalty arrangement which would assure it of two essential things: one, no royalty payments during years of loss (a royalty "holiday," in other words); two, a built-in guaranteed annual rate of return on its investment. And this led the consortium to support a formula for royalties based on earnings, the so-called "net profits" approach in which, instead of paying a flat percentage on the value of all production, the corporation pays a share of its net profits, i.e., profits after allowed deductions, to the government. To say the least, such an arrangement does not exactly encourage profit-making zeal in a company; indeed, it can create precisely an opposite tax incentive for a corporation to transfer profits "downstream" into other areas of its vertically integrated organization. There is in fact an incentive to avoid sharing profits with the government; moreover, in years when the company is showing losses the original owners of the resource gain not a single penny. Under Syncrude's fifty-fifty profit sharing arrangement, for instance, Great Canadian Oil Sands, a company which until 1974 had shown no profit since commencing production in 1967, would have paid no royalties whatsoever to the Alberta treasury. Tied to this "new, innovative and imaginative" approach, as Peter Lougheed later described it, Syncrude was also asking that it be guaranteed a flat return on its investment, a base of eight or ten percent of its investment which it would recoup every year, rain or shine. How could this be written into the agreement? Quite simply, by allowing Syncrude to deduct, say, eight percent interest on its investment as an allowed cost before arriving at the net profit figure to be split with the government. Not only, therefore, was Syncrude demanding to be freed of any obligation to give back a percentage of production to the owners of the resource in years when it was not showing profits; the consortium was also requiring the state to guarantee it a minimum yearly rate of return. And to top it off, already in the spring of 1973 Alberta's tax experts were holding out the promise of tax relief in case Ottawa refused to permit Syncrude to deduct the amounts paid to the province under the net profits arrangement from its federal taxable income. But then, as we are so often informed, oil is a very "risky" business.

Instead of rejecting outright this transparent attempt to have the public underwrite Syncrude's risks, the Lougheed government now took the peculiar step of asking that the royalty issue be given for comment to the Canadian Petroleum Association, official lobby and mouthpiece of the large foreign oil companies operating in Canada. Given the CPA's predictable views and its special interest in the question, this was rather like asking the goats for advice on how to tend the cabbage patch. On June 4, 1973, CPA officials met with members of the Lougheed cabinet and handed in their impartial verdict on the royalty dispute. The government's formula was unsatisfactory, the public should underwrite some of the project's risks, and the royalty should be based on earnings and protection provided for periods when these were absent. In other words, the net profits approach should be adopted.

The incident was revealing of the government's general approach to bargaining with the oil industry. As the Syncrude negotiations moved toward their final phase in the summer of 1973, Lougheed's predicament was a political one: how to keep face and to avoid the appearance of a blatant sell-out while giving the Syncrude group its essential demands. Peter Lougheed had himself concluded from his reading of Walter Levy's report and his energy discussions in Sweden that the tar sands must be brought into production as quickly as possible if Alberta hoped to cash in on the growing energy crisis abroad. All his hopes rested on the Syncrude application; the project must not be permitted to fall through. The premier understood that the companies participating in Syncrude would demand a stiff price for a "go" decision, but in return he would need some assistance in selling the deal to the public. For a long time he had preached the need for "participation" in the tar sands by "citizens of Alberta," and this, it was clear, would have to figure in the Syncrude deal. The solution, devised by Syncrude's negotiators and the government's public participation committee, was for Alberta to have a minority, passive ownership in the project. A fifth company, the Alberta Energy Company, would be created to participate financially in the Syncrude plant and its auxiliary facilities. Syncrude was more than willing to meet Lougheed's political needs: why not, the consortium too has an interest in appeasing Canadian nationalist sentiments, and the government was only requiring a twenty percent option in the plant—a far cry from the participation requirements the same companies are facing from the major oil producing nations. The company would be run as private enterprise's answer to the solutions of the New Democratic Party, with government putting up fifty percent of the equity and private investors the remainder. Government involvement in the plant would thus only amount to ten percent. But in addition, the government was intending to have only a passive, financial interest in the Alberta Energy Company, taking up a minority of the directorships, so there would be little potential threat to the corporate sector here. As Syncrude clearly understood, Lougheed was protecting his political base, designing an instrument which would not offend "free enterprisers" but which would satisfy his need to demonstrate that the people of Alberta—at least the few who own stocks—could benefit from tar sands development. Inevitably, the company as constituted will function as a political pressure group, backing rapid development, high prices and large returns. If its share value drops, the government will come under immediate pressure from the directors of the company to make additional concessions to the companies developing the tar sands. And so, paradoxically, this type of public company will sustain the momentum for more and more similar resource deals. It is shrewd stategy and Syncrude knows it. So long as the consortium is permitted to protect its other leases, retain control of development and exercise the right to license its technology, it has an interest in supporting a degree of passive state participation. This issue later became crucial in Syncrude's later dispute with the federal government.

The negotiations were due to be concluded by the end of August 1973. On June 18 Syncrude's negotiating team, which always included representatives of the four participants plus Spragins and his staff, met with the Lougheed cabinet and deputy ministers. Lougheed had asked to "hear the whole story" and Syncrude was glad to oblige. The companies spelled out a long list of fourteen concerns and the action that would be required by various government departments for a decision by August 31. The list of concerns—over and above royalty and public participation—included assurances of labour stability, environmental regulation, housing priorities in Fort

McMurray, material and labour availability, and so on. In effect, the company was tying an entire package of risks, costs and commitments to its agreement to give a go decision. Within six weeks the government had essentially complied with Syncrude's terms. Early in August Mines and Minerals Minister Bill Dickie informed Syncrude that all government departments involved had taken the necessary steps to set in motion the vast provincial infrastructure which the companies would need in the construction and operating stages of the project. The infrastructure would include housing priority for Syncrude, the paving of Highway 63 linking Edmonton and McMurray and other roads, the initial public cost of which Dickie estimated in excess of $60 million. Since then this figure has mushroomed to nearly $300 million!

Two other concessions carried no price tag, although both reduced Syncrude's risks substantially. In mid July Syncrude was given important commitments relating to its future position under government regulation of the environment. The minister of environment, Bill Yurko, spelled out the Department of Environment's requirements, enclosed permits required under the Clean Air Act and Clean Water Act for Syncrude to commence construction, and then gave the consortium two additional assurances. First, should the government anticipate any changes in standards of corporate performance expected under the Clean Air and Clean Water acts during the construction phase of Syncrude's project, it would promise to discuss such changes in detail with Syncrude before enacting any new rules. Second, when Syncrude applied for licenses to operate its plant under the same acts, the licenses would be issued for a full five-year period; that is, the licenses would establish the conditions under which the plant would be allowed to operate for five years. These two assurances appear to reduce Syncrude's uncertainties regarding expensive changes in environmental regulation, and they obviously also limit the authority of the Department of Environment to enforce tighter standards, particularly in the crucial first five years of the project's life. Ecologists will deplore the fact, but environmental standards in the tar sands are already being decided in part by political considerations. In Syncrude's case, the government's desire to see the development of lease No. 17 go ahead apparently led it to sacrifice some of the flexibility it will need to regulate the largely unknown environmental hazards of large-scale tar sands projects.

A second concession, later written into the agreement as one of three Syncrude conditions, related to the consortium's demand that the government somehow guarantee no strike, no lockout labour conditions while the plant was being built. Syncrude wanted a tight assurance from Lougheed's cabinet that it would enact legislation if necessary to accommodate the participants, one of whom (Atlantic Richfield) had made guaranteed labour stability a contingency of its decision to proceed. What Syncrude was demanding was difficult, for existing legislation prohibited the kind of separated "site agreements" the oil companies wanted, and construction contractors in the province were fiercely opposed to a site agreement as well, arguing that the high price which would be paid to the construction unions would trigger similar demands on other projects. The government nonetheless promised to use all of its resources to get all parties to support a voluntary site agreement. If the unions resisted, the government would "re-examine" its policy toward collective bargaining in the construction industry, and if this failed, Syncrude was told, the government would step in to legislate what the oil companies needed. This commitment essentially delegated to a foreign-owned group of companies the government's legislative authority in this area. Within a year, on Syncrude's demand, the government duly legislated.

Syncrude's ability to set the terms and conditions of tar sands development had by now been convincingly demonstrated. The government had yielded to virtually every fresh demand, absorbing the oil corporations' risks and costs and placing itself in the humiliating posture of a supplicant literally begging Syncrude to proceed. Peter Lougheed would attempt to rescue the initiative with one last bluff, but it was far too late to reverse the roles. The multinational corporation had already dictated the rules of the game, government would simply have to camouflage its own lack of authority and power. In the unequal contest between the oil cartel and the Lougheed government, the superior strength of Syncrude had been decisively established.

9   The Politics of Syncrude
I I: Politicians on the Hook

'The central consideration emerging from the analysis of the politics of oil is the incompatibility of a socially irresponsible system of power with the goal of a truly democratic society. A corrosion of democratic principles and practices pervades wherever the interests of private oil and public policy meet. The attractions of privilege and power derived from the control of oil invites pressures, from small as well as big business, that mock the ideals of responsible government, a just society, and a peaceful world.''

Robert Engler,
The Politics of Oil: Private Power
and Democratic Directions

Canada's politicians have mortgaged the future of the tar sands—twice—and tried to camouflage the fact. In September 1973 Peter Lougheed seriously misled the people of Alberta and the rest of Canada when he assured them that the first round of negotiations with the Syncrude consortium had been settled "on the government's terms." Incontrovertible documentary evidence demonstrates precisely the opposite: the negotiations were settled on the oil companies' terms. A year and a half later, following the withdrawal of one of the original partners, Syncrude imposed a new set of terms and an ultimatum, and once again Canada's politicians capitulated. This chapter examines the evidence surrounding the earlier giveaway of 1973; the two following chapters recount and attempt to explain the threatened collapse of the project and its rescue by three Canadian governments over the winter of 1974-75.

Syncrude's own documents reveal that in July of 1973, with the

August 31 deadline for a "go" decision only weeks away, the Lougheed cabinet began to have some eleventh hour reservations about the oil companies' terms. At issue, as in the later confrontation with Ottawa and Alberta, was Syncrude's insistence that since developing the tar sands would be a marginal economic proposition at best, the public sector must absorb and underwrite some of the project's heaviest costs. Previously, the government had evidently agreed with this thesis. Now, however, two unanticipated developments fed suspicion that the oil companies might be playing a devious game. First, there was the matter of Syncrude's costs. Originally fixed at about $500 million, the consortium's cost estimates had been overtaken by a phenomenal rate of inflation, hitting $650 million in January 1973, $750 million by March, and in July, when the last revisions were handed to the cabinet royalty committee, just under $1 billion. What this suggested was that the oil companies were artificially inflating Syncrude's cost estimates in a bid to win additional concessions from government and to take advantage of the profit sharing arrangement. Under profit sharing it is obviously in the interests of the participants to exaggerate their costs and overcharge Syncrude through the use of high management fees, royalty fees, rental payments, licensing arrangements with the parent firms, and so on. Through such techniques, essentially intra-company financial manipulations which all the oil companies long ago mastered, Syncrude's "profits" can be kept to a minimum, re-appearing elsewhere within the corporate structures of the owning companies.

Secondly, the price of crude oil had been on a steady upward trend for almost a year. Four successive increases had boosted crude prices by almost a full dollar a barrel.[4] It was increasingly a seller's market in oil, a fact which put the tar sands in a new light. In view of the rising price trend, was the future of this vast resource quite so "marginal" as Syncrude was claiming?

Syncrude's accountants and the government's own financial advisors, Foster Economic Consultants of Calgary, a subsidiary of a U.S. consulting firm, had worked closely over the months, using alternative royalty schemes to project various rates of return. They had worked so closely, indeed, that at one point Syncrude was offering the government consultants assistance in drafting royalty rates! Foster and Syncrude used the accounting method employed by all the oil companies in planning their investments: discounted cash flow (DCF) analysis allows a company to find the optimal rate of return on its investment over a certain period of time. DCF analysis is equivalent to finding that interest rate which, if paid by a bank on the original investment and compounded over the project's life, would yield at the end the same amount of dollars as the projected investment. The oil companies generally regard a DCF rate of return of at least ten to fifteen percent as the minimum cut off point on relatively risk-free investments. In its negotiations with the Lougheed government Syncrude argued that its DCF return was marginal, anticipated costs were high and increasing, and risks were heavy; thus Syncrude should be charged no royalties during years of loss and should be permitted to deduct all its costs and losses carried forward, plus eight percent interest on its investment before having to share net profits. With all of this the Lougheed cabinet had agreed. But now, quite unexpectedly and at the last minute, the government attempted to reverse the roles and impose new terms. The ensuing confrontation was brief but sharp, and when the smoke lifted the mastery of the oil cartel had been decisively re-asserted.

For several successive days in late July and early August 1973, Peter Lougheed's cabinet met to consider its "final terms" on royalties and other outstanding issues. Syncrude's negotiating team was standing by in Edmonton, waiting for the wrap up discussions. Then on the afternoon of Friday, August 4, Mines and Minerals Minister Bill Dickie telephoned Syncrude to say that no further meetings would be necessary, a letter was on its way setting out the government's conditions. Moreover, the cabinet had made a final decision; the terms would not be negotiable.

Minutes later the letter was delivered to Syncrude's offices. In it Dickie had spelled out the various concessions and commitments the government was willing to make to accommodate Syncrude's lengthy list of concerns. Then he added the "final" cabinet guidelines on royalty, public participation and lease renewal. The last two posed no problem. Lease renewal would now require an exploration commitment by the companies to assess the value of the tar sands lying beneath each lease, but this presented no threat to the monopoly of the large companies over the key leases. Syncrude's participants could continue to hold one sixth of the leased acreage in the core surface mineable portion of the Athabasca deposit. As previously agreed, citizens of Alberta would have an opportunity to buy shares in Syncrude through a public corporation holding eighty percent of the pipeline, fifty percent of the utility plant and a maximum twenty percent option in the Syncrude plant itself. On the key issue of royalties, the government was still committed to profit sharing instead of ordinary royalties and was offering relatively generous terms: Syncrude could deduct from gross revenues (1) all its operating costs; (2) normal "straight-line" depreciation costs; and (3) losses carried forward in determining net profits. However, the agreed guaranteed rate of return in the form of interest on investment was nowhere to be seen. Moreover, the government had inserted an insurance clause giving it the option of taking a seven and one-half percent ordinary royalty on production after five years, and in exceptional circumstances the whole royalty question could be reviewed after five years.

The government had unilaterally revised the royalty terms, dropping the lucrative, unearned base return on investment. With several hundred million guaranteed dollars abruptly snatched from their pile of chips, Syncrude's negotiators were furious and determined to call the cabinet's bluff. Dickie was contacted again and an immediate face to face encounter was demanded. The minister was reluctant—it was by now late Friday afternoon and he was not eager for a confrontation with Syncrude, but finally he relented.

And so eight oil men—one representative from each of the four participants and four members of Syncrude's staff—raced over to Bill Dickie's office for what one of them aptly called "the showdown." Frank Spragins warned Dickie that the omission of the guaranteed rate of return could kill the project; perhaps it was an oversight? No, said Dickie, the omission was deliberate, the cabinet had made its decision and the next move was up to the participants. According to the oil men the new royalty formula "doubled front-end costs," it was totally unacceptable, and Syncrude could not recommend the project to the four owners unless the base rate of return was re-offered. Each of the executives took his turn berating the government's terms and threatening immediate withdrawal from the project. The companies would state their final conditions and the government would be given a few days in which to reply. If a favourable response was not forthcoming, then the companies would abandon the project and make a prompt "no go" public announcement—an action sure to embarrass the Lougheed government. Dickie was asked by the spokesman for Atlantic Richfield to remind his cabinet colleagues "that they would not find a group willing to spend a billion dollars very often. They would not find another customer should Syncrude decline. The government had been playing poker with Syncrude for a year and this was the showdown." Alberta's minister of Manpower and Labour, Dr. Bert Hohol, joined the meeting and was promptly informed that the consortium no longer regarded his assurances on labour stability as reliable enough. It would be a mistake to leave the tar sands under a system of volun-

136

tary arbitration. Syncrude wanted watertight guarantees of government intervention if a no strike agreement was not achieved in 1974.

Syncrude President Frank Spragins explained in a letter to the four participants, dated August 8, 1973, what happened following the encounter with the hapless Dickie:

> "Discussions continued when possible during the week-end. A meeting was held with Dr. Hohol . . . at which a more conclusive statement on labor was worked out. . . . Also, considerable discussion took place with Dr. Hohol regarding survival of the Syncrude project. He appeared to show considerable concern and volunteered to discuss the matter with the Premier and, if possible, to set up a meeting with the Premier to be attended by Syncrude representatives. Dr. Hohol was able to reach the Premier by phone over the week-end but reported that he was unable to arrange a meeting or convince the Premier of the extreme seriousness of the situation. In the meantime, Mr. Dickie had reported the developments to Mr. Getty [Don Getty, minister of Intergovernmental Affairs at the time] and in turn Mr. Getty was in touch with the Premier. Early Tuesday morning Mr. Getty called Syncrude and said that if the government had made any mistakes regarding the Syncrude project then the Cabinet would be prepared to hear our arguments. Accordingly, a meeting was arranged whereby the Syncrude group would meet with appropriate members of Cabinet at 2:00 P.M. on August 15th at which time Syncrude's response to the government's letter of August 3rd would be discussed.
>
> "In addition Mr. Lougheed has requested that the senior executive officers of the participants meet with him at 11:00 A.M. on August 16th . . . ."

Meeting in Toronto on August 9 to thrash out a joint strategy, Syncrude's management committee agreed to prepare its final counter-offer by August 14 and to present it to Alberta on a take it or leave it basis. If the government insisted on royalty revisions, then it must offer other concessions which would yield the same overall rate of return. The common front of the cartel would be held as well: it was agreed "that Shell Oil be approached to state to government that it shares Syncrude's concerns in regard to the economics of the project."

The threat could hardly have been blunter. The government "would not find a group willing to spend a billion dollars very often. They would not find another customer should Syncrude decline."

The message went straight to the heart of Lougheed's concern that the tar sands would never be developed if Syncrude's project should fall through. In effect, Syncrude was threatening to use its veto power to block development of the tar sands. Here were the holders of the leases coolly informing the supposed owners of the resource that the tar sands would only be developed if and when the oil companies decided, and then strictly on their terms. This was equivalent to an apartment tenant writing the terms of his tenancy and presenting them to the owner of the apartment building in the form of an ultimatum. The difference, of course, lay in the fact that the large oil companies monopolize the tar sands leases, the lease potential information, technology and expertise, and were able to exploit this monopoly power to call the government's bluff once again.

In theory, of course, the cabinet might have countered with its own ultimatum. It could have said, for instance, "Then we shall invite another lease holder to go ahead on our terms." But the government was already bargaining with the entire oligopolistic industry; Shell and the other majors were kept informed of the Syncrude negotiations and were unlikely to break the common front. The government might also have said, "We shall ask someone else, say the Germans or Japanese, to develop the tar sands in cooperation with Canadians." Or it could have decided to go the public development route, taking full responsibility itself for the tar sands. But the government had already ruled out the public ownership option, giving up its strongest and ultimate bargaining card because it was politically impossible and ideologically unthinkable; and to invite anyone else into the tar sands would mean tearing up the leases or forcing the majors to give them up. The government's real choices, then, were either to give in to Syncrude's terms or to challenge once and for all the monopoly power of the lease holders. Don Getty's telephone call ended any remaining doubt about which side would determine the conditions of tar sands development. The government's hasty retreat from its so-called non-negotiable position meant that the tar sands were effectively under the control of the oil cartel and that the government held only the shadow, not the substance of sovereign power in the area of Alberta's resources. The public relations veneer thrown over the sell-out would camouflage the capitulation—Lougheed would even tell the public that the negotiations had been settled "on the government's terms"—but from now on there would be no further dispute about who wrote the rules and who was "on the hook."

What transpired at the later meetings between Syncrude and the Alberta government can be deduced by contrasting the cabinet's "final terms" of early August with the actual agreement signed on September 14, 1973. The agreement was a conditional one, a letter of intent, and amounted to a tentative "go" decision. The significant changes appearing in the September version occurred under the heading "Provisions Concerning Royalty." In calculating the net profit to be shared with government, Syncrude would now be permitted to deduct (1) operating costs, (2) depreciation, (3) losses carried forward, and (4) eight percent interest annually on seventy-five percent of its overall investment. In addition, the consortium would be allowed eight percent interest on its construction costs, up to a maximum of $90 million. These last two interest allowances, amounting under Foster Consultant's original estimates to a total of $989 million over the project's twenty-five year life, constituted the guaranteed return on investment which Syncrude had so vehemently demanded. This sum would be deducted from net profits; thus the loss to Alberta was half of $989 million, or nearly half a billion dollars. Well worth a week's intensive lobbying!

But this was by no means all. Syncrude had also attached three conditions to its "go" decision. First, labour stability must be assured. Second, a federal tax ruling would have to be given allowing Syncrude to deduct all royalties (as a share of net profits) paid to Alberta under the profit sharing agreement from its federal taxable income. Finally, the third condition: "that the federal government does not regulate directly or indirectly the prices of synthetic crude oil below the levels attainable in a free international market." The last two conditions on tax and pricing meant that Syncrude's lobbying machine was about to turn its attention toward Ottawa. What were the implications of the oil companies' new conditions?

Syncrude was asking for special tax treatment on the same grounds it had used so successfully with Lougheed, namely its "marginal" economic status. The consortium was well aware that a 1971 federal tax ruling would not permit provincial royalties based on income (for example, profit sharing) to be deducted from income tax after 1976. It was demanding exemption from this ruling. Under Foster's original calculations more than $1 billion over twenty-five years would be written off Syncrude's taxable income if this special tax treatment were extended. Under present estimates, depending on how well the profit sharing formula is enforced by Alberta, the consortium could be able to write off a good deal more than this.

Syncrude's third condition—that the consortium have guaranteed access to world prices—was inserted in reaction to Ottawa's imposition of a temporary crude oil price freeze and an export tax on oil shipped to U.S. refineries. What Syncrude wants is unrestricted access to international prices, the right to charge what "the market" will bear. It argues that its real competition will come from oil from shale and coal, from the future fuels of the U.S., thus the market in these commodities must be allowed to determine the price of tar sands synthetic crude. But if U.S. producers have their way, oil from U.S. synthetic fuel industries will be both heavily subsidized and extremely expensive; moreover, they will be guaranteed a domestic market and protection against cheap imports. Syncrude was demanding, in effect, to be included in a continental pricing system in which the estimated price per barrel of future fuels will be well in excess of $10. Shell Canada Limited, in fact, has maintained that tar sands oil must be "very much higher" than $10 per barrel—a purely arbitrary demand which appears to be utterly unrelated to real costs. The lease holders in the tar sands are saying that if Canada wants this great resource developed, then it must be prepared to pay incredibly inflated prices or allow the producers to export to the U.S. market. Canadians will either pay the international price for future fuels or they will get no oil from the tar sands.

In lobbying for its tax and pricing concessions in Ottawa the Syncrude group had the staunch political backing of the Lougheed government. It is quite clear that Peter Lougheed had made a two-part deal with the oil companies. First, he would use his position and influence to pressure Ottawa. Second, if this failed, his own government would rewrite its agreement with Syncrude. Small wonder that Syncrude considered the Alberta premier in their debt. "Premier Lougheed called today," Spragins informed the participants on September 21, a few days after the slick television presentation of the deal, "and personally congratulated and thanked Syncrude on the manner in which we handled the confidentiality of the negotiations with the government, and the manner in which Syncrude handled its end of the publicity with regard to the announcement to proceed." Lougheed said he had written to Prime Minister Pierre Trudeau explaining the tax and pricing conditions attached to the agreement and outlining the reasons why the project should go ahead. Lougheed advised Syncrude to prepare an economic evaluation of the project for use in Ottawa and promised to "be in touch with Mr. Armstrong on Monday for further discussion on strategy." "Mr. Armstrong," no doubt, was J. A. Armstrong, then president of Imperial Oil. Peter Lougheed and Exxon, with tigers in their tanks.

Lougheed was literally acting as Syncrude's political strategist, advising the group on tactics, setting up meetings and trying to smooth the way to Ottawa. His public posture was pugnacious and menacing, the row over the export tax was at its height. Privately, however, Lougheed was fronting for the oil companies in Ottawa in an attempt to avoid even heavier concessions by Alberta to keep Syncrude on the rails. It was not the first or last time that Lougheed would front for the petroleum industry: he and Saskatchewan Premier Allan Blakeney had both mastered the rhetoric of provincial rights and western alienation and would soon use it to force prices up, a policy strongly supported by the industry.

Early in October 1973 Lougheed briefed the federal minister of Energy, Mines and Resources, Donald Macdonald, on the tar sands project; and in Ottawa Syncrude found a sympathetic audience in eighteen top federal bureaucrats from various interested departments. Lougheed then advised Syncrude to have the chief executive officers of the four partners meet with Trudeau, Macdonald and John Turner, federal Finance minister. Syncrude wondered if Lougheed was trying to avoid ''some of his commitments to Syncrude regarding a solution to our Ottawa problems. In view of Lougheed's current difficulties in discussing anything with Ottawa officials, the proposed strategy may not be too far out of line,'' Spragins informed his masters on October 12. ''However, before embarking on this plan of action, it should be understood that this does not let Lougheed off the hook in case Syncrude representatives are not successful in Ottawa.'' If Syncrude failed to win its tax and pricing demands in Ottawa, the companies intended to take a hard line with Alberta to force Lougheed to devise a fresh formula yielding the required rate of return for the project.

Syncrude's lobbying machine was also working hard to influence opposition political parties, business and professional groups and the press. In early October 1973 Spragins and other Syncrude officials briefed Conservative party leader Robert Stanfield three times in as many days during a Tory tour of Alberta. At first Stanfield seemed bored and uninformed, but after visiting the tar sands for a day he ''made a far better accounting of himself,'' according to Spragins.

''At this stage Mr. Stanfield had spent a day visiting Fort McMurray and the tar sands, and had become an 'expert.' His recent press conference has indicated at least some improvement in his understanding of the energy problem—especially from Alberta's point of view.''

Thereafter, question period in the federal House of Commons saw frequent exchanges on the Syncrude project between the government front bench and Alberta Conservative MPs such as Peter Bawden, Eldon Wooliams, Harvie Andrie and others. The oil lobby speaks with many tongues but the message is always the same.

Frank Spragins, Syncrude's able chief lobbyist, was also paying careful attention to the job of keeping the public informed about the "realities" of tar sands development. Smarting from accusations that Lougheed's terms were actually a sell-out, Spragins arranged (with the enthusiastic backing of the provincial cabinet) to give a lengthy speech in Edmonton in November correcting Syncrude's critics. At around the same time, two revealing interviews with the ex-Mississippian ("a Canadian by choice") appeared in the Toronto *Globe and Mail*. Spragins, remarked the interviewer, "can be quite lyrical when talking about the proposed plant"; it embodied "a personal and industry goal of having synthetic oil on tap as an alternative fuel source, with production from twenty-five or more similar facilities, when most of the conventional oil wells of North America will have run dry." But Syncrude was still negotiating with "its back door always open," Spragins warned. If the federal government was unwilling to give special treatment to the project, then almost certainly "his team of specialists and their synthetic production know-how would be turned to the rival Rocky Mountain oil shales in the United States." The tar sands had once been well ahead of the oil shales, but massive U.S. financial aid could see the Alberta resource overtaken as a future fuel source. "In the United States, the government gives all sorts of incentives to develop synthetic fuels production technology; in Canada, we are surrounded by negative attitudes and discouraging measures." But Syncrude—this giant "all-Canadian endeavour"—was not asking for handouts, just "fewer disincentives." The petroleum industry was "thinking in terms of thirty to forty years from investment to recouping of outlays with a decent return," and Syncrude would certainly be building other plants if its first panned out. The oil industry claimed, however, that Canada must act quickly to avoid losing out to the oil shales.

Shades of Walter Levy! Here was the oil cartel back at its favourite political pastime, playing alternative producing regions against one another in the endless search for better payouts and lower risks. Once again Syncrude was exploiting its position as the spokesman for the major leaseholders, threatening to hold back tar sands development if its terms were not met. But how real was this threat?

At no point did either the provincial or federal government seriously investigate Syncrude's threat to "by-pass" the tar sands in favour of other future fuels. Had they bothered to do so, they might have discovered some interesting facts. They would have learned that the majors and the U.S. government regard the oil shales as an essentially unproved resource, as a giant question mark. Shale development has proceeded very slowly under a prototype leasing program designed to weigh the ecological effects and commercial feasibility of an oil shale industry.[5] The leasing program has encountered fierce resistance from environmentalists who argue that shale mining will result in air and water pollution, in disruption of wildlife by an influx of men and machines and the denuding of the surface, and in sociological disturbances as thousands of workers and their families move into the sparsely inhabited shale areas of Colorado, Utah and Wyoming. Environmental groups such as the Sierra Club have been threatening legal action to halt shale development on public land and to tighten up the ecological conditions of the leases. By contrast, although the environmental hazards of tar sands production are at least as severe, there has been relatively little organized opposition by ecologists in Alberta to the strategy of rapid, forced development. Another restriction on large-scale shale exploitation is the lack of sufficient water for the mining and extraction process. Project Independence envisages large production from shale, but according to *The Wall Street Journal* (July 24, 1974): "Developing the oil shale industry to its potential level of three to five million barrels a day . . . would require using all the available water in the Colorado, Utah and Wyoming region, where the shale exists. This, of course, would have vast implications for the region's agriculture, its future population growth and its overall economy." The companies also face far more expensive leasing terms in the shales than in the tar sands, where the leases go for an insignificant annual fee of twenty-five cents per acre. In the shale leases, the majors must first submit competitive bids for a large tract of shale land; after that a developer pays an annual rent of fifty cents per acre plus production royalties based on the grade of shale mined. The winning bid on one of the first public tracts went for $42,000 per acre!

The likelihood of coal liquefaction processes overtaking the tar sands is even more remote, and it is not at all clear in what sense nuclear power can render nonconventional hydrocarbons such as tar sands synthetic fuel a "useless" or "obsolete" asset, since the two fuels are not substitutional and are unlikely to be in direct competition with each other. Tar sands oil is of such quality that it would certainly be in high demand in Canada, especially as a petrochemical feedstock. In fact, the longer one considers the whole bogey of "bypassing" the tar sands, the flimsier the argument becomes: the lease holders are engaging in an elaborate bluff constructed on nothing more than a foundation of empty threats and clichés. All government would have to do to call the bluff would be to change its policy on renewal of the leases, most of which will expire over the next decade. Perhaps expecting just such a move, Syncrude protected its position by including lease renewal provisions in its package agreement with Peter Lougheed.

The Liberal government of Pierre Trudeau, no less gullible than Peter Lougheed's, took Syncrude's bait and began to worry that the fate of the tar sands really did rest in the hands of one consortium. Syncrude's terms must be met. At the end of October federal Energy Minister Macdonald and Finance Minister Turner met with their Alberta counterparts and proved to be sympathetic to Syncrude's tax and pricing demands. Turner agreed to meet Syncrude's tax needs, but the concession must not be in conflict with federal policy that after 1976 royalties based on income would no longer be deductible from taxable income. So a loophole would have to be fashioned.

A way out was found by having Alberta and Syncrude rewrite their agreement, deleting any reference to royalties and substituting the concept of a joint venture. Turner was then prepared to rule that for Syncrude's federal income tax calculations, Alberta's share of the net profits of the joint venture would not be regarded as taxable income: this billion-dollar promise would also have "the strongest protection" that could be given Syncrude. However, Turner was anxious that the Liberal government not appear to be a party to the Syncrude agreement, so he wanted all reference to federal actions deleted from the amended version. (Although the Syncrude group did not know it, Turner's assurances would later grow in significance. As we shall see, in 1974 the issue of taxation of royalty payments reared its head, but Syncrude escaped the burden of heavier taxation because of these earlier commitments from Ottawa.)

Syncrude's insistence that it be given a written guarantee of unrestricted access to high prices was more politically awkward for the Liberals. Domestic oil prices had been temporarily frozen in September 1973 and international prices had begun to skyrocket shortly thereafter. By year's end OPEC had effectively quadrupled the world price of crude oil, pushing the average price to between eleven and twelve dollars per barrel; some oil went for as much as $18 per barrel. At home the oil companies and the producing provinces wanted the freeze ended and prices boosted. Industry wanted the entire Canadian market to reflect world prices, arguing that much better profit margins would be needed to develop frontier oil and gas and the oil sands. The multinational oil companies have been insisting that future investment in resource development must be paid for out of retained earnings (internally generated income being the great insulating strength of the modern corporation) and this requires far higher rates of return. In its annual report for 1973, Imperial Oil argued that "the costs of finding, developing, producing and transporting frontier hydrocarbons will be far greater than the costs for prairie oil and gas. The costs of mining the tar sands, extracting their oil, then replacing the sand, are going to be high. At Cold Lake, the underground rock formations containing the heavy oil must be heated to make their oil thin enough to flow to the wells—a very costly process. Making coal environmentally acceptable as well as easily transportable involves large capital investments. All these things can be done, but they point inevitably to higher costs—very much higher than prairie oil. There is no avoiding this unpleasant fact: the days of cheap energy are over."

Now the Liberals and their mandarins essentially accepted the oil company line—the Department of Energy, Mines and Resources had been arguing for some time that the only real constraint on energy supplies was price—but the minority government's dependence on the New Democratic Party, then holding the parliamentary balance of power, restricted its ability to manoeuvre. Threatened by the NDP with defeat in the House of Commons in early December 1973, the Liberals extended the price freeze on domestically produced oil and announced the creation of a national petroleum company, moves which were generally interpreted as severe rebuffs to the oil industry. In their preoccupation with the short-term concessions to political expediency, however, both the NDP and the media missed the real message of Prime Minister Trudeau's December 6 speech. Much of the speech dealt with the need for immediate action to develop frontier and unconventional sources of energy supply in Canada. Trudeau expressed cabinet desire for the early building of the Mackenzie Valley natural gas pipeline, even though the proposal had yet to be examined by the regulatory agencies. Then he got down to the economic facts of life. "A characteristic common to all frontier oil, and oil from unconventional sources," the prime minister informed the House of Commons, "is that it is much more expensive to develop and to bring into production than wells in the southern part of Canada." This meant "that the days of abundant, cheap energy for Canadians must come to an end. It means that the cost of producing domestic oil must go up before too long. It means that Canadians must be prepared sooner or later, in one way or another to pay for these additional costs or go without oil . . . . In short we must in the long run allow the price of domestically produced oil to rise toward a level high enough to ensure development of the Alberta oil sands and other Canadian resources but not one bit higher.''

But what was ''high enough,'' where would the line at ''not one bit higher'' be drawn? According to the government's own data in *An Energy Policy for Canada,* at least 15 billion barrels of tar sands oil could be available for five dollars a barrel while 35 billion barrels would be recoverable at a price of six dollars per barrel (1972 dollars; the estimates include, of course, a decent return to the producer). But these prices have already been exceeded: within months of Trudeau's speech, GCOS synthetic fuel was selling for more than seven dollars a barrel and the oil companies were calling for prices of ''very much higher'' than $10 a barrel!

Trudeau's speech was evidently intended to signal industry that its demand for higher prices and better returns would be partially met. If 1973 was a record year for oil profits, 1974 would be better yet as Canadian prices followed international prices up the escalator. On December 10, 1973, four days after Trudeau's statement, Donald Macdonald wrote to the Syncrude principals concerning their conditions on international prices and access to export markets. Macdonald's letter quite correctly refused to give specific commitments tying the hands of future governments, particularly in view of ''the multiplicity of future and unpredictable events which may occur.'' However, the Energy minister went on, ''I am able to state our general policy with respect to Alberta oil sands production, as follows:

''(1) To encourage investors in the private sector, and the government of Alberta, to undertake the development of the Alberta oil sands in the most expeditious way so as to ensure the continuity of domestic supply and to maintain a reasonable level of exports;

''(2) By allowing prices for oil in Canada to be based on international prices over the longer term, provided that such international prices are fair and reasonable to the economy of Canada and to the interest of Canadians generally;

''(3) To ensure that the full production of the Alberta oil sands may be marketed with a first claim on such production by the Canadian consumer and thereafter, to the extent that we have the reasonably foreseeable available supplies additional to those necessary for use in Canada, then to export any such surpluses to Canadian requirements or other foreign purchasers at just and reasonable prices in such export markets.''

If Macdonald's qualified statement of intent was not as specific as Syncrude wished, it did go part of the way to meeting certain of the consortium's terms. The letter implied that the federal government intended to encourage a policy of rapid tar sands development, one tied to the option of exporting surplus oil from the resource. It ruled out any ''shutting-in'' of tar sands production by guaranteeing that ''full production . . . may be marketed.'' And it repeated Ottawa's intention that over the long run Canadian oil prices would be based again on international prices, though this is qualified by the nebulous ''fair and reasonable'' criterion. But the real message that came shining through these paragraphs was this: Ottawa's policy towards the tar sands would be almost indistinguishable from its traditional policies of conventional oil and gas development. More of the same, but for higher prices.

Syncrude had one more score to settle in this round of negotiations. A last condition attached to the September 1973 agreement would have to be met by Peter Lougheed's cabinet before a go-ahead would be assured. Syncrude had insisted throughout the 1973 discussions that it would require a no strike, no lockout agreement with the construction unions building the plant. If this could not be accomplished through voluntary agreement, Syncrude had insisted, then the government must legislate amendments to its labour laws permitting separate ''site agreements'' in the tar sands. By early 1974 matters had reached a stalemate because of opposition and obstruction from other construction employers to talks between Bechtel and the unions for a no strike pact. Bechtel and the unions had been forced to break off contract discussions after the Alberta Construction Labour Relations Association threatened them with legal action. Fearing an increasingly vulnerable situation with a growing work force and no guarantees against stoppages, Syncrude applied new pressure on the provincial authorities to legislate an end to the impasse. In the summer of 1974 Alberta complied by enacting laws which permitted no strike arrangements in the tar sands. But the real author of the legislation was Syncrude. Lougheed was merely implementing another of the promises he had given the consortium during the 1973 talks.

Thus were Syncrude's terms for tar sands development won by the constant flexing of its political muscle. By any standard the list of incentives and concessions was a long one: a guaranteed base return;

major tax write-offs; royalty holidays and a profit sharing formula to ease front-end costs; promises of labour stability; commitments of heavy public funding for infrastructure, and so on. But Peter Lougheed was satisfied that the price had been worth paying: the incentives had initiated the momentum of tar sands development, and by early 1974 Alberta's cabinet ministers were confidently predicting that many a plant would follow Syncrude. Ambitious plans for an oil sands corridor were being drawn up for northeast Alberta, the province's policy of economic diversification had been launched by the commencement of the giant Syncrude venture. Nothing, it seemed, could puncture the buoyant mood of the province; if anything, the real danger lay in too much growth, too soon.

But Alberta's boom could also quickly become a bust if the oil industry challenged Lougheed's policies and power. New events were unfolding in Canada's energy crisis—events far removed from the tar sands—which saw Alberta, Ottawa and the major petroleum companies involved in a far-reaching struggle for control of Canadian resources. By the winter of 1974-75 this struggle reached a climax in a two-month test of wills which began with the threatened collapse of Syncrude and ended in a Winnipeg hotel room.

10   The Politics of Syncrude
III: The Illusions of Oil Power

"For which of you, intending to build a tower, sitteth not down first, and counteth the cost, whether he have sufficient to finish it?"

St. *Luke 14:28*

"The meeting did not end until about 8 P.M., meaning Mr. Macdonald, Mr. Chrétien and the others from Ottawa had to find a place to spend the night. But the International Inn had no rooms. Perhaps when the great Canadian movie is made it will include a scene depicting how the federal team, the details of a brand new $2 billion deal locked in their briefcases, trudged down the wintry highway near Winnipeg International Airport in search of shelter for the night. With not a suitcase or a toothbrush among them . . . .

Another hotel about a mile away took them in and they headed downtown to celebrate at a Chinese restaurant named, appropriately enough, the Shangri-la (according to the dictionary, a remote, beautiful, imaginary place where life approaches perfection)."

(Geoffrey Stevens,
Globe and Mail,
February 6, 1975)

Of such gripping stuff is Canada's political and economic history made. What was being celebrated over chopsticks at the Shangri-la, of course, was the shotgun marriage of three U.S.-owned oil companies and three Canadian governments in a frantic eleventh hour rescue of the giant Syncrude project. Behind closed doors in a

149

Winnipeg hotel room, in little more than the average Canadian's workday, a handful of oil men, politicians and bureaucrats had played out the last hand in what Canadian journalists were calling the great $2 billion Syncrude poker game. By day's close the politicians had committed better than a billion dollars of public monies in direct response to an ultimatum issued by the companies two weeks earlier, Syncrude had three new participants, and Canada had turned another corner of the energy crisis. Depending on one's point of view, the moment was either full of tense political drama or staged farce— perhaps a bit of both. How did it happen? Were the governments bluffed and blackmailed into huge concessions to the companies? What are the broader implications of the Winnipeg deal?

In a nutshell, it was a test of wills, an exercise in power politics. What happened to the Syncrude project between December 1974 and February 1975 cannot be understood unless the whole affair is seen as part and parcel of, first, Canada's prolonged dispute with the international oil industry over domestic prices and taxes, and, second, the acrimonious political divisions these same issues have created inside the country. Syncrude was made the fulcrum for a test of strength among the principal actors in the conflict for control over the disposition of Canada's developed and future energy resources. To be sure, crucial questions of economics were involved. At stake, as in Peter Lougheed's earlier ''showdown'' with the same companies, were the financial rules of the game, the terms and conditions under which the nation's existing reserves of oil and gas and its unexploited energy would become available. But more than this, the large dominant companies were flexing their political muscle, testing their power to influence events and shape policies with threats to create economic chaos and withhold supplies of vital resources. Years ago Lenin summed it up in a famous maxim. ''Push your bayonet as long as it meets mush; when it meets steel, withdraw!''

Little but mush stood between the oil companies and the winning of their objectives. By exploiting their monopoly veto power—their power as exclusive holders of essential information, expertise and technology to scuttle development of the tar sands—the corporations were able to push up the price of new energy development in Canada while establishing a milestone on the road to higher energy prices and better profits. What confronted this demonstration of corporate strength was a near vacuum of public policy. Operating in an atmosphere of fear over energy scarcity and without the most basic knowledge of the issues at stake, Canada's feuding politicians chose the line of least resistance, harnessing the resources of their governments to the interests of the multinational corporations.

There was, it must be emphasized, nothing accidental about such an outcome. Surrender was built into the crisis from the outset by Canada's long-standing dependence, a dependence which translated into a loss of leverage and an absence of policy. In an independent, self-reliant nation the Winnipeg meeting would not have happened. Power requires a psychological medium of submissiveness and fear if it is to thrive: threats and blackmail work only when the target, be it an individual or a people, has been conditioned to react like a victim—to react, that is, with fear and without a sense of autonomy or self-confidence. Power also thrives in a situation in which one side in a conflict controls the other's information and its sources of intelligence concerning the issues at stake: if knowledge is power, ignorance is weakness. And finally, power succeeds when one party to a dispute can exploit the divisions and break the unity of the other side, practicing the ancient policy of *divide et impera,* divide and rule. In international politics these Achilles heels—psychological reliance on outsiders, ignorance of one's own capabilities, internal disunity—are symptomatic of dependent status, a situation in which a weaker nation is locked in unequal economic, political and cultural relations to a stronger country. The terms of these relations are arranged to benefit the more powerful nation and its junior allies in the dependent state. In Canada's case, our national dependence programs resource sell-outs of the type negotiated over Syncrude.

Let us now retrace our steps and attempt to locate the origins of the second Syncrude crisis. One useful reference point would be the decision of Canada's first ministers in late March 1974 to raise the price of domestic oil from $3.80 to $6.50 a barrel and to hold it at that level until July 1975. This consensus was worked out after much political haggling and months of bitter federal-provincial dispute over Ottawa's announcement in September 1973 of a temporary freeze on Canadian oil prices and the imposition of a tax on oil exported to the U.S. These moves, announced as part of an anti-inflation package, provoked accusations by Alberta's Peter Lougheed that the federal government was engaging in a "power play" for control of the province's resources.

Provincial rights aside, we should remind ourselves that petroleum prices had been regulated only after a year of sharp successive increases by Imperial Oil and the other integrated majors—increases ordered by Exxon Corporation whose sole justification lay in the prices then prevailing in the U.S. export market. Thanks to Alberta's low royalties, Ottawa's even lower income taxes and the high level of foreign ownership in the oil industry, Canadians were actually losing money on each price rise. But the oil companies were anticipating unprecedented windfall profits. "The strong market demand and the recent series of crude oil price increases, followed as they are by rapid product price jumps," noted *The Financial Post* a month before Ottawa moved, "will bring the fattest profit rises in history this year to all segments of Canada's petroleum industry." But the frequency of the price rises, cautioned the newspaper, was building up "increasing consumer pressure for establishment of some form of two-price system that would protect the home user from the full impact of these and further expected rises."

Although forewarned, the petroleum industry had been predictably hostile to Ottawa's regulation of prices. Lamented *Oilweek,* it marked "the first real major government intervention in oil marketing in Canada since the early 1960s when the National Oil Policy was established." Commencing the industry's lobbying campaign to roll back this alarming new precedent, W. O. Twaits, then chairman of Imperial Oil, warned: "To peg crude prices artificially will put future supplies in jeopardy by inhibiting exploration and development, particularly in frontier regions, and rendering exploitation of the tar sands uneconomic." In plain language, either the federal government would quickly end the price freeze or industry would withhold vital energy supplies. To drive the point home, Imperial and its partners in the Syncrude project made it a last minute condition of their agreement to proceed that synthetic oil be permitted access to export markets and international prices. But, as we have seen, Syncrude did not get its unconditional guarantee on prices from Donald Macdonald; nor did the temporary price freeze come off as promised in early 1974.

As world oil prices and Canada's contentious export tax soared upward, the winter of 1973-74 saw the Conservative government in Alberta, the NDP regime in Saskatchewan and the foreign-owned oil industry forming an unofficial front against the NDP-backed minority Liberal government in Ottawa: the energy crisis makes for strange bedfellows! For the oil companies this tacit alliance was shrewd strategy since it tied their vested interests to the popular motherhood issues of provincial rights. A prairie farmer might not have great sympathy for Imperial Oil's plight, but could surely be counted on to support policies which promised to fight eastern exploitation and indifference. While the oil industry turned its propaganda resources, which are considerable, to the task of persuading consumers that the age of cheap energy was indeed over, the producing provinces demanded "fair market value"—i.e., world prices—for their fast depleting nonrenewable resources. The lines and tone of the debate seriously obscured some of the vital energy issues confronting the country as a whole; and the media compounded the confusion and mystification by dwelling almost exclusively on the war of words between Edmonton and Ottawa, much of which was sheer rhetorical wind. In the absence of well-defined national policies on energy linking the common interests of all regions of Canada, the initiative would by default be left to the majors; which was precisely what the oil industry intended.

What the industry did not intend, however, was that the dispute between Ottawa and Edmonton would turn into a fight over revenue sharing. Having grabbed virtually all of the 1972-73 price increases (which totalled 95¢ a barrel), the large companies recognized that they would have to share any additional windfalls with the producing provinces. The Lougheed government had had to struggle hard against strong resistance from the industry to increase Alberta's oil royalties from about sixteen to twenty-two percent, but this conflict predated Ottawa's regulation of prices and the proclamation of the export tax. Once the federal government began to tax Albertan oil destined for the American market, Lougheed quickly freed himself of the constraints of inflexible fixed royalties and decreed that henceforth they would float up with prices.

Behind the new Alberta royalty concept, which was accepted by the industry as inevitable if unwelcome, lay the ambitious blueprints for the forced industrialization of Alberta then being sketched out by the Conservatives and their empire-building allies. These men saw the world's energy crisis as the perfect opportunity to use the province's remaining energy as leverage for transforming an oil-based economy into one in which resources would be upgraded all the way to plastics through petrochemical processes. But to be competitive and in order to attract the globe's chemical giants to Alberta, the government would have to subsidize the factories with hundreds of millions of public dollars—much as it was underwriting Syncrude's costs by providing a guaranteed return on investment to the four oil companies in the project. The need for large public subsidies for industrial diversification (plus the need to protect himself from political criticism in Alberta for having yielded all proceeds from the export tax to Ottawa) led Peter Lougheed to overreach his actual political power and sow the seeds of a crisis which simmered for months and finally erupted in the Syncrude affair.

The decision of Canada's first ministers to raise oil prices to $6.50 a barrel as of April 1, 1974, represented a major advance for the petroleum industry and its allies in Alberta and Saskatchewan. During the federal-provincial bargaining which predated the pricing accord, however, Ottawa had served notice that it intended to have a slice of the surplus $1.9 billion the new price would produce. Two weeks prior to the meeting of first ministers, Lougheed was given a pointed warning to this effect by Prime Minister Trudeau: "I must make clear," wrote Trudeau, "that any action that you may decide to take in respect of royalties would have to be without prejudice to our freedom of action as regards federal taxation." Undaunted, Lougheed returned to Alberta with the new oil price and promptly announced a new schedule of royalties: the existing average royalty of twenty-two percent would apply to the first $3.80 of the price of a barrel of oil; on the increase of $2.70, however, the province would take sixty-five percent on "old oil" (oil already in production) and thirty-five percent on "new oil" (oil from new discoveries, extensions of existing wells or enhanced recovery techniques). This left the oil industry an extra dollar a barrel on oil in production and considerably more than this on new oil. Canadian Petroleum Association spokesmen complained loudly about Alberta's new levies, but the industry was looking at an extra $740 million from its share of the new price. Then Ottawa moved.

On May 6, 1974, the oil war escalated again with the appearance of the federal budget. Finance Minister John Turner's House of Commons statement contained several unpleasant surprises for the industry-Alberta-Saskatchewan group. In addition to toughening the depletion allowance and reducing exploration write-offs significantly, Turner announced that provincial royalties would no longer be deductible in the calculation of federal income tax. If enacted, these proposals would have increased the petroleum industry's annual tax bill by as much as $630 million.[6] The budget thus would have squeezed the industry, but the real objective of Ottawa's action was political: to force the provinces to retreat by creating a major rift in their unspoken alliance with the oil companies. The May budget was never implemented—the NDP forced a new election—but Turner and Trudeau made it clear during the campaign that the non-deductibility of royalties was firm Liberal policy (they also confirmed Syncrude's specific exemption from the policy). Most of the major oil companies announced cancellation of exploration plans in a none-too-subtle attempt to influence the voters, and Trudeau was confronted by hostile demonstrations during a campaign visit to Calgary. The demonstrators, it turned out, were mostly oil company employees who had been encouraged to take a holiday and hassle the prime minister. But the return of the Liberals with a comfortable majority on July 8 meant that the Lougheed administration was now exposed to the full brunt of the industry's wrath. There could be little doubt as to the outcome.

Alberta derives over forty percent of its economic productivity from the oil and gas industries, and, as the companies never tire of reminding the province, one in three Albertans depends indirectly or directly on these same industries for his or her livelihood. Politically, this dependence on one foreign-controlled industry leaves the province extraordinarily vulnerable to pressure from the few companies which account for most of its royalties and which control the fate of the many Alberta-based "independents"—exploration, service, pipeline companies, etc. Imperial Oil alone accounts for 20.4 percent—one fifth!—of Canadian oil production; seven companies account for more than sixty percent, ten companies for more than seventy percent. These, the affiliates of the largest international oil firms, also control much of Alberta's leased oil and gas properties, they own all refining, control most of the nation's pipelines and have a stranglehold on the marketing of their own products. The budgetary decisions of these corporations are thus of urgent public interest; the oil industry's power to wage economic warfare on the people of Canada, its ability to sway governments through its entrenched hold over vital resources, have long since ceased to be private concerns.

Peter Lougheed began to mend his fences with the oil industry in July 1974 when his government quietly reclassified 10 trillion cubic feet of natural gas from "old" to "new" gas, thus reducing the supplementary royalty from sixty-five to thirty-five percent. The Canadian Petroleum Association had been urging such a step, worth potentially hundreds of millions to the industry, for months; it was taken, incidentally, against advice from the Alberta civil service that the province's "incredibly generous" royalties in the past should rule out new concessions. But the reclassification of gas reserves did not resolve the issue of oil prices and how the increased revenues should be distributed among the three parties to the triangular dispute. Back and forth between Ottawa and Edmonton, playing their own version of shuttle diplomacy, the industry's chief lobbyists kept up their campaign to force the two levels of government to roll back royalties and income taxes. During National Energy Board hearings held during the spring of 1974 on the question of Canada's oil exports, all the major companies presented gloomy projections of the country's short-term supply position and contended that much better rates of return would be required to provide the necessary cash flow for exploration and development of new energy reserves, including the tar sands.

Hitherto, the tar sands had not been regarded as significant to Canada's immediate energy requirements: announcing his agreement with Syncrude in September 1973, Peter Lougheed had commented that much of the plant's product would have to be sold in the export market. By the next spring, however, Syncrude's publicists were arguing that the project and its successors would be vital to Canada's own supply needs. Without rapid development of the tar sands, the Energy Board was warned, Canada would not only miss the target of energy self-sufficiency; by the 1980s she would be dangerously dependent on costly, insecure offshore oil. Unless Canadians came up with between $35 and $50 billion of extra revenues from oil already discovered—say, $5 or $6 net on each barrel of the estimated 8 billion barrels of recoverable oil—and also provided guarantees of high returns on tomorrow's oil, then the tar sands and frontier energy resources would remain undeveloped. Scarcity had thus become the cutting edge of the oil industry's game plan.

As John Turner and Donald Macdonald subsequently confirmed, shortly before the federal budget of November 18 was announced Ottawa reached an understanding with the big oil companies. The nature of the deal, Macdonald revealed, was that if Ottawa introduced a less punitive version of the May 6 budget, "the companies would turn their attention to the provinces." And this is precisely what happened. Turner backed off his May taxes by about $100 million, leaving industry with just under thirty percent of oil production income. But royalties were declared non-deductible, and the budget implied that the industry's share of income would decrease as Canadian oil prices rose—unless the provinces retreated on royalties. Not surprisingly, the industry renewed its shrieks of "double taxation" and several large companies—Imperial, Gulf, Pacific Petroleum, Chevron Standard, Home Oil—simultaneously announced major reductions in planned spending for oil and gas exploration in 1975. From Calgary came daily scare stories of drilling rigs heading for the United States; local gossip had a few of the smaller companies ordering their drivers to move oil field equipment only after dark, and the mayor of that city denounced the two governments, "a plague on both your houses," for jeopardizing the future of his metropolis. Across Alberta, just a few months earlier apparently the most fortunate of Canada's provinces, descended the unfamiliar pall of instant recession. "If I wasn't such a charitable person," reflected Donald Macdonald, "I'd say it's a threat of blackmail over the budget issue."

Peter Lougheed has never been a relaxed, spontaneous politician, but he had seldom looked so strained, so unsure of himself, in his public appearances. "His soul seems as clenched as his jaw line," Christina Newman wrote shortly thereafter, "and the prime impression he leaves is of a man under great tension, handling it with implacable resolution." Comparing the style of Quebec's sixties rage against Confederation with Alberta's demands for rights for the West, a former senior civil servant from Ontario told Newman:

"The Quebeckers postured in the Gallic manner but they'd relax afterwards over a drink. These Alberta guys are grim around the clock. What they're after is not language rights or an end to economic disparity. They come from a rich province and what they want are two things I'm not sure anybody can give them. . . .

"They want redress for the slights they think they've suffered and they want absolute power within their own turf. If you take a look at Lougheed himself or any of the interchangeable four or five guys who come with him to any meetings, you'll believe me when I say we're in for dangerous times."

Of course, to an Ontario mandarin any threat to the imbalanced *status quo* in Confederation could be viewed as "dangerous times." Moreover, the anonymous civil servant was mistaking Lougheed's style for the substance of power. By the end of 1974 the Alberta premier's blunt oil weapon had been revealed as something of a paper tiger. In the spring he had badly miscalculated Ottawa's resolve to have a share of windfall petroleum revenues; he had overreached his capacity to tax an industry powerful enough to bring Alberta's economy to a standstill; he had lost his argument "in principle" with the federal export tax; he had been frustrated in his effort to block Sarnia's big Petrosar ethylene development, a project he and his ambitious business allies viewed as inimical to their industrial plans; and there were signs that all the grandiose hopes for rapid exploitation of the oil sands were evaporating. Lougheed's much touted "oil power" was mostly illusion.

But Ottawa too was under pressure. If Alberta's political weakness lay in its exposure to economic coercion, the federal government's was to be found in a growing anxiety over domestic energy supplies and Canada's relations with the United States. Hard choices were in the offing. In October the National Energy Board reported to the Trudeau cabinet on the contentious issue of oil exports. The NEB document, "In the Matter of the Exportation of Oil," confirmed that Canada was indeed facing the threat of petroleum shortages in the near future: between 1975 and 1982 Canadian oil "producibility"[7] would decline from 2.1 million barrels a day to 1.51 million barrels daily. Over the same period the nation's demand for its own oil (i.e., demand in the market west of the Ottawa Valley, plus 250,000 barrels a day for Montreal after extension of the Interprovincial Pipeline) would grow from about 1.2 million barrels a day to 1.53 million barrels a day. The NEB's projections showed demand outstripping supply in a scant 7.3 years: an outlook grim enough "to require expeditious action to reduce crude oil exports." Although the loss of net self-sufficiency at an early date was unavoidable, the board rejected proposals for a rapid termination of oil exports on several grounds. Such action would only extend the period during which producibility would exceed demand for Canadian oil by two-and-a-half years, noted the NEB, but it would also "have a very serious impact upon exploration programs and the development of additional productive capacity," reducing additions to reserves by up to fifty percent (no evidence was provided to support this somewhat arbitrary conclusion; historically, Alberta's reserves were developed in conditions in which production was shut-in substantially). On the other hand, the board dismissed industry's wish to continue exporting oil at the very high rates which had prevailed since 1972-73: "continued full exports in the face of an approaching shortage would not be in the best interest of all Canadians." So, as a halfway house, the NEB proposed to adopt a new protection formula regulating exports and to commence a seven-year phasing out of oil exports in January 1975. "However," it warned, "if there is no significant increase in the producibility levels now forecast, there will be a rapid phase out of all exports." These recommendations were accepted by the Liberal government in November as the basis of its new export policy for oil.

Several points must be made about the NEB's highly pessimistic supply/demand projections. First, as noted, no account is taken on the demand side of the potential impact of conservation policies: the demand lines are based on the assumption that requirements for indigenous crude oil will grow at an average annual rate of 3.2 percent over the next twenty years. Clearly, the Canadian government

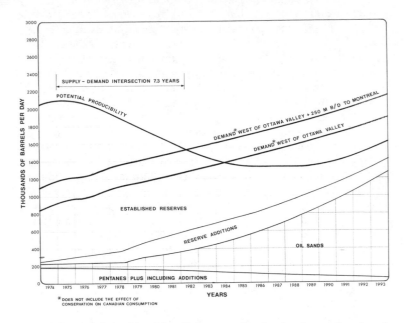

NEB FORECAST OF CANADIAN CRUDE OIL
AND EQUIVALENT SUPPLY AND DEMAND
FOR INDIGENOUS FEEDSTOCKS

should be tackling this side of the equation by implementing tough policies which discourage consumption, encourage conservation but do not distribute the costs of such action inequitably: industry's solution, simply to let oil and gas prices rise several-fold, is a modern version of a policy popular among Russia's tsars—"The shortages will be distributed among the peasants."

Second, the NEB's "potential producibility" line may turn out to be too conservative, at least over the long run, since it discounts the availability of oil from the Arctic. Moreover, several oil companies have suggested that improved recovery techniques could make substantial additions to the established reserves of the western sedimentary basin given, of course, the proper price, royalty and tax conditions.[8] Remarkably, in light of its past performance, the NEB's estimates are generally more bearish than the oil industry's.

Finally, the NEB's projections illustrate the growing importance of the oil sands in Canada's supply picture. The board assumed one Syncrude-type extraction plant going onstream every second year

159

after 1978, with two *in situ* projects coming into production as well in the late 1980s. Oil sands production would be above 500,000 barrels a day in 1985, around the million mark in 1990, if the NEB estimates panned out. Though far less rosy than its earlier wild forecasts of oil sands potential and more cautious than the projections of the large companies and Alberta's Energy Resources Conservation Board, this was still an optimistic timetable. One plant every two years would be a monumental undertaking, putting severe strains on labour supply, materials and smaller businesses; and without putting the economy on some sort of extraordinary footing, the target of the National Energy Board simply could not be met.

But suppose no further development of the oil sands occurred; that Great Canadian Oil Sands was left isolated in the wilderness, turning out its small output of crude, and that no other plants were built. Then not only would Canada's long-term supply position and its future balance of payments begin to look exceedingly bleak, failing rapid development of Arctic oil discoveries, but the anticipated shortfall of domestic oil would appear as early as 1979 and necessitate a faster reduction in oil exports to the United States than that envisaged in the NEB report. If the option of developing the oil sands disappeared, in other words, Canada would face severe supply and balance of payments difficulties by the 1980s and increased friction with the U.S. almost immediately.

Both the federal and Alberta governments, then, were under stress in late 1974 because of the changing configurations of the energy crisis. Ottawa had successfully demonstrated its ability to tax the industry and to drive a wedge between the oil companies and the producing provinces, but it was vulnerable on the issue of future energy supplies and its trading relations with the United States. The Lougheed administration had badly overreached its power to defy the central government, and now its entire strategy of building a western industrial base was being threatened by the oil industry's slow squeeze. Exasperated by more than a year of governmental regulation and intervention in what it had grown to regard as a private preserve, the petroleum industry was fighting back on all fronts and growing daily more determined to force a confrontation, a test of strength, with Ottawa and Alberta on the future course of Canadian energy development.

11 The Politics of Syncrude IV: Veto Power

BARBARA FRUM: *"Is this called being over a barrel?"*
DONALD MACDONALD: *"Well, ha, ha, ha, being over, I suppose, 125,000 barrels."*

"As It Happens," CBC Radio
January 30, 1975

By the first week of December 1974, while a Canadian government delegation headed by Prime Minister Trudeau travelled to Washington to explain the implications of its new oil export policy to the Ford administration, Alberta was in the midst of a political crisis over the re-introduction of John Turner's resource budget. Intense public and private pressure was being applied against the provincial leadership by the industry to force Alberta to reduce its royalties and taxes, Calgary was in an uproar, companies were pulling out or threatening to do so, and an alarmed media had swung solidly behind the industry's demands. Against this backdrop Alberta's ambitious plans for the tar sands suddenly came apart.

Announcing the inability of his government to devise an oil sands development policy in light of uncertainty surrounding Ottawa's intentions on pricing and exports, Peter Lougheed told the Alberta legislature in late October that federal energy decisions threatened the province's timetable for exploiting the sands. Shell Explorer, a wholly-owned affiliate of Shell U.S., was pulling out of its fifty-fifty venture with Shell Canada because there was no assurance that it would be able to export its share of oil production to American refineries. Escalating construction and operating costs were also cited by Shell Explorer as reasons for its withdrawal. In November, shortly after the announcement of the federal budget and the release to the press of the National Energy Board's report on oil exports, both the Athabasca Oil Sands Project group and Home Oil-Alminex Limited let it be known that their projects too were being put into cold storage. At the same time Alberta Mines and Minerals Minister Bill Dickie worried aloud that Syncrude's estimated costs had soared above $1.5 billion and could reach "well over $2 billion."

On December 4 the crucial blow was delivered. Atlantic Richfield, one of four participants in Syncrude, tersely announced its immediate withdrawal from the giant project—better than a year into construction with about 1500 workers on site north of Fort McMurray. The decision vividly underlined Alberta's vulnerability and its lack of autonomous decision making power. "The shock waves generated sent some of North America's leading oil executives scurrying to a meeting in New York as well as making employees in Fort McMurray and Edmonton worry about their future," noted the Edmonton *Journal,* a newspaper which has consistently supported Lougheed's tar sands policy. ARCO's spokesman in Washington denied that a recent refusal by the United States Export-Import Bank to lend the consortium $75 million for the purchase of U.S.-made equipment had influenced its decision: U.S. Treasury officials, however, hinted that the Export-Import Bank's decision (actually taken by the National Advisory Council, an inter-agency group headed by Treasury) had been motivated in part by a concern that American energy supply be protected. ARCO explained that its action had been determined by rising cost projections which had become "just unbelievable": the firm had reached "a point where we have to give something up."

In point of fact, ARCO had been disenchanted with Syncrude for several months. Rumours that the company was thinking of quitting the tar sands were circulated in Fort McMurray in the summer of 1974. Faced with rapidly escalating costs in Alaska, where it holds some 2 billion barrels of north slope oil and over twenty percent of the Alyeska pipeline, the company had been experiencing serious cash flow problems in meeting its commitments. Though it has a good chance of improving its overall position within the industry once Alaskan oil goes to market, ARCO's ambitious expansion program, emphasizing diversification into petrochemicals and alternate energy sources, had begun to strain its internal and outside capital resources. As part of a general strategy of rationalizing its holdings and easing the financial burden, the firm withdrew from the first major project in the Colorado oil shales in September 1974, and it is reasonable to assume that its decision to abandon Syncrude was made around that time as well. Federal announcements on the phasing out of exports probably reinforced the inclination to quit: the company has no refining capacity in Canada and the parent firm is already a heavy net purchaser of crude. The timing of the announcement, however, appears to have been arranged to have maximum impact on both levels of government—one of which saw the Syncrude project as vital to its future economic plans, the other of which was depending on Syncrude to help relieve the nation's deteriorating oil supply position. From the point of view of the majors ARCO's action not only increased the financial burden of the other members of the consortium; it also offered the possibility of using the project as political leverage in bargaining for some old and new concessions. Syncrude was strategically placed, the perfect candidate for the industry's test of strength with Canada's political leadership.

For the Lougheed administration the threatened collapse of Syncrude posed grave political consequences. Without Syncrude there would be no further exploitation of the tar sands and many of the province's plans for diversifying the economy would have to be shelved. So much of the economy and the Conservative government's prestige had already been tied to Syncrude's coattails that the sudden failure of the venture would throw a couple of thousand workers out of their jobs, send many small companies to the wall and puncture the buoyant atmosphere which had prevailed in the province to that point. Such an event could also have far-reaching political repercussions, perhaps even arresting Social Credit's steep slide into oblivion—a feat somewhat akin to refloating the *Titanic*—and certainly improving the prospects of the provincial New Democratic Party which was calling for development of the tar sands through a publicly-owned crown corporation. Peter Lougheed had deliberately created public expectations of growth and spectacular progress and his personal image and credibility were now bound up with Syncrude's fate. He had burned his bridges: the project would have to be rescued.

On December 12 Lougheed eased his crisis with the oil industry by implementing his Petroleum Exploration Plan, a package of tax credits and royalty reductions designed to return between $300 million and $500 million a year to the allegedly overburdened industry. The plan had been fashioned during the summer of 1974 as a response to Turner's May budget proposals and, according to a version which was leaked to the press, had the objective of holding the oil industry's life-cycle discounted rate of return near its very healthy current level of fifteen or sixteen percent. The leaked document confirmed that without higher taxes the industry could be expected to reap tremendous profits on the order of $60 billion or more by century's end. Nonetheless, Lougheed's PEP concessions went beyond the earlier contingency plan, restoring industry's share of net production income to thirty-three percent and leaving the oil companies with a realized income (at the $6.50 per barrel price) of $2.38 per barrel of "old" Alberta oil, $2.95 on "new," after payment of royalties, taxes and lifting costs. The plan was publicized as beneficial to the small independents, and it was, yet the inevitable effect of reducing royalties and taxes was also to improve the profitability of the largest operators. As the province's big producers, Imperial, Amoco, Shell, Gulf and Texaco would receive much of the benefit of the concessions. Ottawa applauded. John Turner had won a unique Canadian triumph.

The terms demanded by the three remaining Syncrude participants, Imperial, Gulf and Cities Service, to keep the project alive were to prove even more expensive. They were made public on January 16, 1975, at a remarkable joint press conference held by the three corporations in Toronto. Taking up an old unsettled grievance, the group demanded access to world prices. Second, Syncrude wanted fresh guarantees of exemption from the non-deductibility provisions of the Turner budget as well as from any future pro-rationing of oil production. And third, the public sector must come up with a billion dollars in equity and/or further tax concessions. All this by January 31, or the project would be closed down. If that happened, noted Gulf Canada president Jerry McAfee, it would be five or ten years at least before any other producer started up another tar sands project. The threat could not have been plainer.

"We'd like to get another billion," remarked McAfee. But was that all the three companies were after? Geoffrey Stevens, *Globe and Mail* columnist, described Syncrude's cross-country lobbying campaign as impressive and pondered, "Is money what the Syncrude partners really want from the two governments? There is a suspicion among federal officials that the money is secondary, that the real object is the security or protection that would come from government involvement in a risky, costly project. With public funds invested, the governments involved would have to make sure taxpayers' interests were not jeopardized for want of such things as price protection, tax breaks and relief from environmental safeguards." Alberta oil man Nick Taylor, who has long since dispensed with the rhetoric of laissez-faire in his quest for the petroleum dollar, calls this the strategy of "marrying the landlord's daughter. You stop paying rent, get the first porkchop and the best bed in the house"—merely by arranging a joint venture with the state. Direct equity participation by government ensures that any antagonisms between the private sector and the owners of a new resource will be blunted by their mutual interest in seeing development succeed, in spite of high prices, environmental problems, labour strife, and so on. And with future world oil prices unpredictable and a new surplus of petroleum combining with slackened demand to weaken OPEC's monopoly position, from the companies' perspective it clearly made sense to hedge their risks by insisting on Ottawa's entry into Syncrude. Privately, Nick Taylor insists, the large oil companies are unanimously in favour of the creation of Canada's national oil company, Petrocan, because it offers the prospect of a bigger porkchop through joint development of frontier energy resources.[9]

"The January 31 deadline is an ominous and odious thing," wrote Geoffrey Stevens in an angry column. "Can any self-respecting government permit itself to be rushed and forced into throwing a huge amount of taxpayers' money on the table just because the other players, a trio of oil companies, are threatening to take their cards and chips and go home? Somehow that smacks more of blackmail than poker." Stevens was not voicing unpopular sentiments. Even the Calgary *Herald,* no enemy of the oil industry, complained:

> "Even Calgarians will be shocked by this attempt to force the hand of their government. . . . The long and short of it is that the Canadian government has been told to change gears by three American oil companies. That simply does not sit well even with those such as the Herald who have argued long and hard against many of Ottawa's energy policies."

Several major daily newspapers across the country, including the Montreal *Star,* the Edmonton *Journal,* and the Toronto *Star* among others, expressed editorial support for some sort of full public ownership or public utility concept in the tar sands; and many prominent Canadians, including Bruce Willson, former head of Canadian Bechtel and Union Gas Company, were demanding outright nationalization of Syncrude. The Toronto *Star* took that position one step further by suggesting the federal government take over Imperial Oil itself! Speaking in Calgary on January 31, hours before the expiration of the ultimatum, B.C. Premier Dave Barrett brought down the house by giving his own answer to Syncrude's demands—"Get out! Get out!"

But Barrett was not among the politicians making the decisions, and the ones who were clearly were not up to the job of turning the crisis into an opportunity. Three essential failings destroyed that possibility and fatally weakened the bargaining power of the governments.

One principal impediment to effective bargaining by the political leaders was their lack of crucial information. When multinational companies and host politicians barter over terms a key element in determining the outcome is control of vital intelligence; historically, the bargaining advantage has been generally held by the international firm, since it possesses and closely protects industrial information, techniques and expertise unavailable to others. Most governments, including those of the developed capitalist world, have lagged far behind the large global companies in their understanding of how the latter are able to exploit the opportunities open to them to manipulate prices, disguise profits, avoid taxes and repatriate capital. Referring to what they term "the managerial dilemma of the nation-state," Richard Barnet and Ronald Müller, authors of the controversial *Global Reach: The Power of the Multinational Corporations*, write:

> "The institutional lag that cripples governments in their efforts to prevent global corporations from circumventing the spirit of tax, securities, and banking laws is due in no small measure to the technological breakthroughs of the accounting industry. The space-age alchemists have discovered the incantations that turn banks into nonbanks, dividends into interest, and profits into losses. . . . Skilled obfuscation is now an essential accounting tool. The challenge is to create a tidy world for investors, regulatory agencies, and tax collectors to scrutinize, which may have little or no resemblance to what an old-fashioned bookkeeper might have called the real world."

In trying to discover "the real world" in their negotiations with the Syncrude partners, the Lougheed and Trudeau governments encountered many key questions they could not answer. By far the most perplexing arose from the huge cost overruns on the project: from a projected cost of half a billion dollars in January 1973, to just above one billion in June 1973, to better than two billion by the end of 1974. Documents were produced by critics to prove that as recently as July 1974 Bechtel was still projecting the ultimate cost of the plant at below $900 million. Cost increases and purported diminishing profits had, as we have seen, played a major part in the earlier round of negotiations with the Lougheed administration; and it was as clear then, as it is now, that the politicians and their advisors simply did not know whose figures to believe. Throughout, Syncrude has consistently produced data on costs which show that the project can at best make a modest return of around ten percent. As world prices skyrocketed, so had the cost projections more than doubled. Coincidence or contrivance? Would companies which are being assured of an internal rate of return of twenty percent on North Sea oil by the British government be investing large sums of capital in the tar sands for half the return? After ten years of study Syncrude had decided to "go" when oil prices were below $4 a barrel; did the companies now need $12 to $16 for the same barrel? Was Bechtel using the project as a test case for its other big construction contracts? Or had the engineering giant totally botched the original projections? Since Bechtel's management fees are apparently arranged on a cost-plus basis, it clearly stood to gain from the escalations.

The significance of the huge overruns lay in the 1973 agreement to substitute profit sharing for the usual royalty on gross production. As capital and operating costs escalated, so too did the interest and depreciation charges which could be written off by the companies under the profit sharing formula. The so-called "base return," subject of Syncrude's showdown with the Lougheed cabinet in August 1973, was now worth some $2 billion over the life of the project. At that rate the province would be fortunate to recoup anything whatever from its profit sharing contract—which is precisely what critics of the agreement had predicted at the outset! Syncrude's spokesmen explained the doubling of costs as a function of inadequate design (which implied that Bechtel's original projections were unreliable) and escalations for heavy equipment, housing, environmental problems; and, in addition, "a contingency allowance, amounting to about twenty-five percent of the cost increase on the plant itself, had been provided to cover unexpected costs including work stoppages, strikes in transportation systems or fabrication plants and other unexpected events." Operating apparently from the premise that one boondoggle justifies a second, Syncrude president Frank Spragins noted that Montreal's Olympic Games had undergone a similar inflation rate. Well might Peter Lougheed have reflected on Mark

Twain's comment on the art of extrapolation in his *Life on the Mississippi:*

"In the space of one hundred and seventy-six years the Lower Mississippi has shortened itself two hundred and forty-two miles. That is an average of a trifle over a mile and a third per year. Therefore, any calm person who is not blind or idiotic, can see that in the Old Oolitic Silurian Period, just a million years ago next November, the Lower Mississippi River was upward of one million three hundred thousand miles long, and stuck out over the Gulf of Mexico like a fishing-rod. And by the same token any person can see that seven hundred and forty-two years from now the Lower Mississippi will be only a mile and three-quarters long, and Cairo and New Orleans will have joined their streets together, and be plodding comfortably along under a single mayor and a mutual board of aldermen. There is something fascinating about science. One gets such wholesale returns of conjecture out of such a trifling investment of fact."

In a bid to improve what Barnet and Müller call "the institutional lag" that divides the corporations and the politicians, the Lougheed government commissioned a series of studies and audits of the project and Syncrude's books by four private consulting companies. In part, the move seems to have been inspired by Lougheed's own concern to cover his administration against political criticism in case Syncrude did go under. Of the four consulting firms, three—Price, Waterhouse Company, Foster Consultants, and Loram International—have close relations with the oil industry. Price, Waterhouse audits the books of many of the large oil companies, including those of the Syncrude partners; Foster is a Washington-based firm which consults for the same corporations; and Loram, connected to Lougheed's old company, Mannix of Calgary, was at the time working on a construction project near Fort McMurray arising out of the Syncrude venture. These were a rather unlikely group to charge with the task of aggressively penetrating the oil companies' monopoly of knowledge. Moreover, it is a well-known fact of business life that consultants tend to tell their clients what the clients wish to hear, and the last thing Peter Lougheed wanted to be told was that he was being fleeced by Syncrude and Bechtel. In light of this, it is rather surprising that Ottawa too was relying on the four studies for information on which to base its decisions. But long before they were in—and none of the reports were received before the weekend of February 1—the politicians had made up their minds to accept the cost figures of the companies and to rescue the troubled project. "A contentious political question since the Winnipeg meeting," wrote *Globe and Mail* columnist Norman Webster on February 11, after the deed was done, "has been the acceptance, by the three investing governments, of the oil companies' cost figures for Syncrude. Apparently a fair amount was really taken on trust." Neither of the two levels of government had the expertise to do anything but, and the four outside consultants only reported after the principal policy decisions had been made.[10] Resolving the information and the knowledge problem is a long-term proposition, and it can only be properly tackled when governments themselves are inside the oil business.

Internal political divisions within Canada also contributed to the fiasco. There is no need to recite the nature of these divisions, they are as old as the country itself. Yet it must be recognized that their existence certainly strengthened the bargaining hand of the oil companies. The remaining Syncrude partners unquestionably have important conflicting interests and objectives among themselves, but, to repeat, the large oil companies usually manage to present a common front when dealing with governments. Their behaviour, to recall the words of the U.S. Federal Trade Commission, "should properly be regarded as cooperative, rather than competitive" in such a situation; they "continually engage in common courses of action for their common benefit." By contrast, Canada's balkanized political system presented the oil companies with no parallel set of common interests, no joint front organized around a shared federal-provincial energy policy. Consequently, each of the governments involved in the rescue operation was pursuing its own interests and running on a different set of tracks. Better than a year of feuding and rhetorical warfare on energy matters had created an atmosphere that was anything but conducive to joint policy making. Even at the summit meeting in Winnipeg the inter-governmental rows erupted, a fact which can hardly have escaped the attention of the corporate executives. About all Canada's feuding politicians could agree to was that it would be preferable to yield to the companies' terms than to see Syncrude shut down. Capitulation was the common denominator which brought the politicians to Winnipeg.

But—and here was the third and decisive weakness in the political setup—the governments involved were literally incapable of seeing alternatives to capitulation. Blinkered by ideological distrust of public ownership proposals and long habituated to relying on the oil industry to make the investment decisions, the politicians closest to the affair—men like Lougheed, Don Getty, Macdonald, and Ontario

Premier William Davis—cut themselves off from the one course of action which might have changed the result. And that was to make a public statement to the effect that Syncrude would not be shut down: that if the oil companies chose to cut and run, they would lose their investment in the project, be sued for breach of contract and shut out of any future development of the tar sands. If required, the tar sands would be developed by crown corporations—but they would be developed. Beyond such a statement of intent, it was open to both levels of government to turn the tables on the oil companies following their extraordinarily offensive press conference of January 16: given strong, effective leadership from the politicians, there can be no doubt that immense public pressure could have been brought to bear on Imperial, Gulf and Cities Service to force them to rethink their position. This would have heightened the tension surrounding the affair, perhaps prolonging the crisis, but it would also have countered the "ominous and odious" tactics of the three companies.

The point is not a philosophical one; it is purely a question of power politics. By renouncing the option of public ownership and development of the tar sands the political leaders lost their sole opportunity to checkmate what we have called the oil industry's monopoly veto power—its power to threaten to block development of resources such as the Athabasca sands. Without such an option in reserve, without a bargaining card of last resort, the politicians simply lacked credibility when they bravely asserted that they would not be intimidated or pressured into concessions. To put it even more categorically, the governments could not have won the showdown with the oil companies without being prepared to nationalize Syncrude and develop the tar sands on their own.

There is not a shred of evidence that any of the governments involved seriously contemplated such action nor is it likely the others would have agreed had one done so. From the outset of the affair the most the politicians envisaged was that other oil companies and/or governments would have to pick up Atlantic Richfield's share while incentives would also be required to keep the other participants in the project. There was little taste among the country's politicians for the kind of thinking which led an Edmonton *Journal* columnist to remark that "true political leadership consists in seeing the potential of historic change, even as the change is only emerging, and of seizing the opportunity to take hold of events and shape them to advantage. We should see the new oil sands problems in that light. We should view them as new opportunities and challenges, to be turned to good account." That was sound advice, but in the Canadian context almost utopian. It seriously exaggerated the ability of a group of politicians operating from a context of dependence to break out of that context. Expectations based on the false assumption that the politicians enjoyed the independence to choose and to manoeuvre freely were doomed to disappointment. Decades of reliance on powerful outside forces—on outside governments, outside ideas and ideologies, outside corporations, outside advice and outside technologies—have bred into Canada's ruling groups, and many of their supporters, a mentality of submissiveness and dependence that effectively obscures alternatives. The politicians yielded the day they decided Imperial Oil was indispensable to the future of Canada.

Ottawa had indicated its willingness in principle to invest in Syncrude shortly after ARCO's departure, and as the January 31 deadline approached each of the companies' demands on taxes, prices and participation was met. John Turner reiterated his pledge that the project would be exempt from the provision in the November 18 budget whereby resource royalties, or payments in lieu of royalties, to governments would no longer be deductible for income tax purposes. Moreover, added the Finance minister in his letter of January 24 to the three companies, this exemption would apply even if Alberta opted for the low seven and one-half percent royalty rather than a share of net profits.

In December 1973 Donald Macdonald had refused to give Syncrude categorical assurances that its product would be sold at the world price. Such a commitment might bind the hands of future governments, especially in light of "the multiplicity of future and unpredictable events which may occur." Now, however, the qualifications were dropped. Syncrude, Ottawa promised, would be permitted to sell its oil at an "internationally related price," adjusted for transportation and quality. At prices prevailing in early 1975 this would probably mean a price of about $13 a barrel: synthetic tar sands oil is of a higher quality than most crudes, thus Ottawa was in effect guaranteeing Syncrude a higher price than the going world figure when the plant comes onstream. Since all Canadian oil prices are likely to rise to international (or at least U.S.) levels by 1979, the commitment may be academic. There is no provision of a guaranteed floor price for Syncrude, but if world prices should decline the oil companies have the added insurance that Ottawa now has a direct interest in seeing the plant make money. The project has also been exempted from any future prorationing—a commitment which could be important years from now if Canada again develops surplus oil reserves.

On Thursday, January 30, the day before Syncrude's deadline, the federal cabinet agreed to invest between $200 and $500 million in the project in order, as Donald Macdonald put it, to keep open the option of oil sands development. That evening Macdonald explained the decision, and the problems he was facing, to CBC interviewer Barbara Frum on the national radio program "As It Happens":

FRUM: "Do you resent being forced to negotiate under a gun like this?"

MACDONALD: "I think I can say that I would prefer not to be in this kind of situation. I think, on the other hand, that the experience in the energy field in the past twelve to eighteen months is that we always seem to be in these situations."

FRUM: "Is it a question of nerve? A lot of people think they wouldn't walk away from the investment they've already got, that this threat that they're going to close down by tomorrow is just that they've got cooler nerves than the government has."

MACDONALD: "They could be right. I'm probably not the best person to judge that."

FRUM: "Are you worried about the kind of clout they have?"

MACDONALD: "Not exactly. The situation is that it's a fact of life I have to deal with. I'd say that my preference would be that we could determine national policies and national priorities apart from the situations of the individual companies, but the fact is that most of the funds available for investment in the petroleum sector in Canada are in the private sector and therefore we have to plan with that as a reality."

FRUM: "So you aren't prepared, then, to let them walk away from this deal. You can't afford to let them walk."

MACDONALD: "I think I'd have to correct you on that. There is too high a price for the Government of Canada. . . ."

FRUM: "One thing . . . that's bothering a lot of Canadians is: how do we know they need a billion dollars? Are we smart enough to look at their books and understand them? Books can be written to say a lot of different things; they can all be true."

MACDONALD: "Yes, I think that's a valid question. We've looked at their books and made a preliminary assessment of what is going to be involved. The principal reason why we're going to be depending upon the independent studies commissioned by Alberta is to confirm or otherwise the conclusions we've arrived at. I think there always is the problem of being able to confirm what the real cost is. . . ."

FRUM: "There's some feeling, I think, on the part of a lot of Canadians, that government does not now have the resources we need to cope with the oil companies."

MACDONALD: "I think that I'd have to acknowledge that traditionally we have been at a disadvantage in the sense that we haven't built up a bureaucracy specifically for the purpose of counselling government on the development of the petroleum industry. One of the purposes in establishing a national petroleum company . . . is to provide to the government the kind of advice from actual operators in the petroleum field, operators who will have the responsibility of doing their best for the Canadian taxpayer. . . ."

FRUM: "But how does that help you make a decision by tomorrow?"

MACDONALD: "It's no help at all by tomorrow."

FRUM: What about deciding making [sic] as one of your terms where that billion is going to be spent? A lot of people are worried that not only is this costing a lot of bucks . . . but that those dollars are being spent on materials and expertise in the United States."

MACDONALD: "Some of the equipment undoubtedly will be imported, the huge draglines for instance. . . . A lot of it will be spent abroad, that's quite true. We try to put some emphasis on as much as possible being spent in Canada."

FRUM: "Mr. Macdonald, forgive me; you seem to be agreeing as I raise all these problems, and yet you also seem to need this deal to go through. Why?"

MACDONALD: "Well, I think the reason we're concerned about having the deal go through is that we could lose the momentum of developing what is a—albeit a high cost—very important petroleum resource in Canada. We believe there is a pretty substantial potential for oil development in Canada, in the Arctic for example, but this can only be established by drilling. There's a question of doubt as to how much oil is there. The oil sands are different in this sense, that we know there is petroleum there and it's really a question of putting into place a system that can bring it out. If this project does not go ahead, then we will have lost an important option in our national energy policy. . . ."

FRUM: "We may be facing a glut of oil at lower prices . . . the tar sands could in fact be overpriced for the market in the future. Are you not worried about that?"

MACDONALD: Yes, this is a concern. I think it has to be a concern about developing Canadian potential, and while we're only talking about investment in one particular plant at this point, we will be faced as we develop higher-cost petroleum resources through the rest of the decade with the continuing concern that if we make a substantial investment to bring on a high-cost source of oil that investment could be prejudiced if there is a surplus developing around the world and therefore the price drops. . . . It's really the trade-off of the possibility of a lower cost, which is not assured, against the greater probability of being self-reliant.''

FRUM: ''Is this called being over a barrel?''

MACDONALD: ''Well, ha, ha, ha, being over, I suppose, 125,000 barrels.''

FRUM: ''Pretty expensive barrels too.''

MACDONALD: ''Very expensive barrels.''

This exchange confirmed that Ottawa perceived Syncrude as a potential key to future domestic oil supply, and that this consideration—and the resulting need to keep open the oil sands option—over-rode all the difficulties and constraints mentioned by the interviewer. This was perhaps a defensible objective; yet it was also open to the government to retain the same option by calling the oil companies' hand and, if necessary, taking over the project. But, as has been argued, the politicians had renounced such measures and thereby acknowledged Syncrude's veto power. Macdonald was in effect confessing that his government lacked the power to respond to the ultimatum of the oil companies: Canada's energy choices were being arranged by foreign corporations.

Ottawa's moves made the Winnipeg summit meeting of February 3 largely anticlimactic. The search for new partners for Syncrude brought the province of Ontario, concerned about future oil supplies and manufacturing contracts linked to the tar sands project, and Shell Canada to Winnipeg along with Imperial, Gulf, Cities Service and delegations from Ottawa and Edmonton. Shell had professed interest in joining the project, but in Winnipeg, according to one account, ''Shell's terms were so high that even the other three oil company executives were surprised.'' Perhaps. It is well to recall that Shell had its own big project pending, that it too was looking for government participation, and that its terms would be vitally influenced by the outcome in Winnipeg. The company was certainly not there to ease the pressure on the politicians; more likely it was there to present a common front with the other large companies. Having failed to impress the governments with his demands (which reportedly included a floor price and additional relief from Alberta) Shell Canada chief C. W. Daniel took his leave. Discussions then turned to the issue of government equity participation in Syncrude and how the total should be divided among Ottawa, Alberta and Ontario. The old feuds quickly surfaced again among the politicians: Peter Lougheed was determined that the minimum for entry must be five percent of equity, Ontario Premier William Davis wanted to make a smaller offer. According to one account,

> "Mr. Lougheed apparently walked in with both guns blazing and laid about with a rudeness that startled many at the table. At one point, when Ontario was talking about investing $50 million in the oil sands project, Mr. Lougheed remarked sarcastically that Mr. Davis should have saved himself the plane fare out, then suggested he take his assistants (actually, two cabinet ministers) to another room to come up with something serious. . . .
>
> "Also not helping things at the Winnipeg meeting was Mr. Lougheed's aggressive attempt to get commitments that Sarnia's petrochemical industry would be discouraged from further expansion. He wanted the commitments as part of the Syncrude package. He got black looks from Ontario and nothing from federal Energy Minister Donald Macdonald."

Alberta's delegates later denied this version of the dispute with Ontario. "Can you imagine our premier saying that?" asked one cabinet minister. The negotiations were "tense and tough, but they were not sarcastic." One can only hope it was so. Of course, it is conceivable that had some of the tough talk rudeness, black looks and posturing been aimed in the direction of the three foreign oil companies, the people of Canada might have derived some tangible benefit over and above the dubious psychic gratification of watching their politicians play the fool while selling out the country's resources.

Under the terms of the "gentlemen's agreement" negotiated in Winnipeg, Syncrude became the offspring of three Canadian governments and three U.S.-controlled oil companies. The new equity shares and capital commitments broke down as follows: Imperial Oil increased its percentage holding by 1.25 percent to 31.25 percent, but increased its dollar commitments to $625 million; Gulf increased its equity share by 6.75 percent to 16.75 percent and its dollar commitment by $235 million. Cities Service decreased its percentage by eight percent to twenty-two percent and increased its dollar commitment by $140 million. Ottawa agreed to come in for fifteen percent, or $300 million on a projected estimate of $2 billion; Alberta took ten percent of the risk capital, or $200 million; and Ontario took five percent or $100 million. The three companies have seventy percent ownership, and, interestingly, Imperial Oil has just enough equity to outvote the three governments combined.

In addition to these new participation arrangements Alberta agreed to loan Gulf and Cities Service $100 million each in the form of debentures which are convertible to equity, and to pick up the entire cost of Syncrude's utility plant (estimated at $300 million) and pipeline ($100 million). But the province is also facing huge infrastructure costs for highways, schools, bridges, services, etc., associated with Syncrude—estimated at $300 million in early 1975. In return for these commitments, the Alberta government left the Winnipeg meeting with ten percent control and the potential of thirty-six percent control (if it converts its debentures and the Alberta Energy Company exercises its twenty percent option in the plant—worth $400 million).

But this does not tell the full story. The three companies also managed to arrange a tax regime for Syncrude which, in effect, means that they will end up paying for far less than seventy percent of the project. Because Syncrude is a joint venture and not a separate company, under federal tax rules the companies can write off their full investment ($1.4 billion) in the project from their taxable income of presently producing oil wells. These write-offs should be worth just over $500 million to the corporations. Second, the companies will be able to claim "earned depletion" allowances at the rate of $1 for every $3 invested in Syncrude—worth another $170 million.[11] The project also draws tax advantages by being classified a "mine" rather than an "oil well." Once it begins producing oil, any payments made to Alberta through revenue sharing or royalties will be deductible from federal taxable income because of Turner's special exemption. Moreover, one-third of the income from the project will escape taxes completely since the governments will be participating via crown agencies not subject to corporate income tax, and this too could end up saving the private partners in the consortium substantial sums of money. All of this is over and above the concessions won by the companies from Alberta in 1973 in devising the profit-sharing formula. All of which bears out the adage, "a good tax accountant is worth a hundred salesmen." And a friendly minister of Finance is worth a great deal more.

From the taxpayer's perspective, then, the celebrations that followed the conclusion of the Winnipeg talks were perhaps a trifle premature. Under the complex arrangements agreed to by the politicians, the public is committed to the tune of some seventy-five percent of the capital costs through equity, loans and foregone taxes, and the people of Alberta must bear the full burden of the related infrastructure charges. In return for these very expensive commitments the public has thirty percent of the ownership, a dubious profit-sharing formula which gives the companies a strong incentive to run the project without a profit, and little or no federal taxes. It is a very high price to pay to keep open the oil sands option: complete public ownership of Syncrude would probably cost far less, since all returns from the project would at least be captured by the public sector.

How much return the taxpayer will recoup under the existing arrangement depends on many variables, including future oil prices, additional cost escalations, the ability of the government representatives on Syncrude's committees to ride hard on the three private partners and Bechtel, how much purchasing of equipment and services is done through in-house transfers within the oil companies, the extent of Canadian involvement in contracting for equipment and services, whether the governments obtain rights to license technology evolving out of the project, how the governments dispose of their thirty percent of the synthetic crude, and so on. On the strength of past performance it is difficult to be optimistic; and it is unlikely that we will ever know the full costs we are bearing because the governments will have an obvious vested interest in concealing the compromises they will be forced to make. Already, the politicians' progress reports on Syncrude are being written with a fog-index dense enough to conceal unpleasant facts and confound sceptics. Unfortunately, Canada has no legislative equivalent of America's new "Freedom of Information Act" which would allow interested citizens access to vital information. What information we are given will be carefully packaged, designed to sell the project to consumers, voters and taxpayers who are footing the bills. Canada's paternalistic political traditions and our poorly developed sense of citizen involvement in national and provincial politics should ensure Syncrude a future far less troubled by close public scrutiny than its past record justifies.

Beyond the specific meaning of the Syncrude sell-out, there lies the larger and more fundamental question of what it implies about the nature of power and democracy in Canada. In his study of "the private government of oil" in the United States *(The Politics of Oil)*, Robert Engler remarks on "the incompatibility of a socially irresponsible system of power with the goal of a truly democratic society. A corrosion of democratic principles and practices pervades wherever the interests of private oil and public policy meet." But the impact of this socially irresponsible system of power is not limited to the United States; this private government of oil has power all over the world. The organized influence of international oil in Canadian society, as seen through the Syncrude episodes, extends to the highest levels of our political life. The American corporation has indeed become a central cog in the national politics of Canada. Thus if private oil is corrupting democratic principles in the United States, in Canada it is also corrupting the possibility of national sovereignty and independence. For the source of the power which lies behind the successful political manoeuvres and pressure tactics of Canada's oil lobby is not to be found within Canada itself; this power is foreign. In the last analysis the politics of Syncrude are the politics of imperialism. What can we do about it?

12   Energy and Dependence

'In terms of Canadian history . . . the people who met in Winnipeg last week merely did what the Canadian governments have frequently done when confronted with national need (in this case energy supplies for the future) and a shortage of capital. They helped.''

The Calgary Herald
February 11, 1975

In spite of the Syncrude rescue operation, the future of the Alberta tar sands remains a major question mark. Rapid intensive exploitation of the giant resource is far more problematic than it appeared during the first anxious days of the energy crisis. And it now seems certain that any further development by the international oil companies will have to be heavily subsidized and underwritten by the Canadian and Albertan governments—unless they are prepared to break the veto power of the lease holders and develop Canada's future energy resources themselves. Neither of these courses of action is likely to be inexpensive or without significant economic and political risk, but it is vital that the option of public ownership of key energy resources at least be seriously debated by Canadians. If this alternative is not kept alive, we may well find that Syncrude has become the prototype for future energy resource development in Canada. How much is self-sufficiency in energy worth to Canadians?

An obvious precedent was established in the 1973-75 bargaining over Syncrude's terms, a precedent that every major resource developer in the country can now use as its base or starting point. Within days of the Winnipeg deal, all other prospective tar sands projects were tied to the Syncrude model of government participation, guaranteed prices, tax concessions and exemptions and trimmed front-end costs. Shell, for instance, announced that it was seeking government investment (as of this writing, Shell Canada is still looking for partners to take up fifty percent of the consortium's risk capital) and some new breaks: "For one thing," C. W. Daniel, Shell Canada president, noted in April 1975, "we would hope for some kind of built-in protection against the possibility of dropping world oil prices, such as are being offered North Sea operators, and are being recommended for U.S. oil shale developers." Shell's mining project is estimated at $2.4 billion, but we can be virtually certain that this figure will rise appreciably as its negotiations progress with Alberta and Ottawa. Home Oil was also fast off the mark to inform both levels of government that its project would remain on the shelf unless it too obtained "treatment at least as favourable as that given Syncrude." There is nothing particularly surprising in such requests; any oil company would be missing an obvious bet if it failed to point to Syncrude as a precedent. Su Oil, owner of Great Canadian Oil Sands, has asked for exemptions from both export controls and the export tax, and it can be expected that access to U.S. refineries for Athabasca synthetic crude will play an important role in future bargaining as the oil companies try to roll back the National Energy Board's oil export restriction policy.

The implications, however, extend well beyond the tar sands. That the world is not facing imminent short-term oil shortages, nor did it in 1973, is now generally understood. Professor Odell remarks in his authoritative *Oil and World Power* that, "the medium- to longer-term outlook for the build-up of significant oil-production potential, not only in areas which are large oil users, but also in areas which could become important new oil exporters, is probably brighter now than they have been at any time in the whole post-1945 period of dependence of the energy-using world on the oil reserves of the Middle Eastern and one or two other countries." These potentials, Odell adds, "undermine the validity of the spectre of a physical shortage of oil in relation to the world's increased needs in the last fifteen years or so of the present century." *The Economist,* which unfashionably predicted in early 1974 that the world was moving into a large physical surplus, noted in March 1975 that "oil was heading for its biggest instant glut." OPEC has been shutting-in some 12 million barrels a day recently, about one-third of its productive capacity, and there have been predictions that this shut-in capacity would rise to 15 or 16 million barrels a day before the end of 1975.

Producing countries such as Libya and Abu Dhabi have already experienced severe problems in selling their oil, and the surplus has manifestly strengthened the bargaining hand of the "seven sisters" of world petroleum. One U.S. energy economist went so far as to argue in the *Wall Street Journal* on March 20, 1975, that:

> "The oil cartel is in the early stages of a breakdown. Crude petroleum prices are being lowered, both directly and indirectly, by individual producing countries seeking to increase their exports. In the next several months, the demand for OPEC-produced petroleum will decline sharply. The cartel will fall apart when its members prove unable to share the necessary production cuts."

Against this extreme viewpoint it can be argued that the current surplus is more a consequence of world recession and reduced demand than of increased alternate supplies, and that OPEC can and will stabilize and even increase oil prices by sharing additional production cuts. Cartels are not permanent, stable entities, but OPEC should be understood as more than a simple producer's cartel motivated merely by economic interests. It is more like a trade union than a cartel, in that shared ideological, psychological and political variables play a major role in its collective decisions, and these may prove to be more durable and binding in the long run than economic considerations. Attempts by the western consuming nations to undermine OPEC are consequently more than likely to increase its solidarity. Moreover, the international oil industry is most unlikely to assist any movement toward the destruction of the present artificial price structure, and there are also important consuming countries that want prices kept high so as to encourage development of their own high-cost energy resources. Nevertheless, the surplus is likely to grow for some time to come, and this is bound to create some strains within OPEC and make it difficult, though certainly not impossible, for the producer's organization to stabilize the world oil market for any predictable time-frame.

Canada's own future energy choices—including its policy toward the Athabasca tar sands—are very much bound up with these unpredictable and unstable world conditions. The nation's shortfalls of domestically-produced crude oil and natural gas, and its locked-in export ties to the United States, may panic Ottawa into pursuing an energy strategy that could leave Canadians paying for high-cost, heavily-subsidized reserves that are over-priced for tomorrow's market. Cognizant of this and the possibility that world prices may decline, the major oil companies will insist that Canadian consumers and taxpayers underwrite much of the risk and cost of developing and protecting new energy resources. The industry's strategy is to shift risks to the public sector while forcing today's consumer to pay for tomorrow's oil, and by mid-1975 there was every indication that the strategy was working. The federal authorities have not only bailed out Syncrude, they have indicated a willingness to raise oil and gas prices substantially, offer low royalties and taxes in the north, guarantee returns, provide expensive publicly-financed infrastructure and so forth. This will apply not only to the tar sands, but to the development of frontier oil and gas reserves, the financing of pipelines, research and development programs for new processes and technologies and in covering the large social and environmental effects of the policy. In such circumstances, Petrocan, the national petroleum company, could conceivably be turned into nothing more than another secretarial and capital pool of the multinational oil industry.

What price is self-sufficiency in energy worth paying? Is it worth twenty Syncrudes? The costs of building new energy capacity in Canada have been estimated by federal officials at well over $100 billion dollars. Should Ottawa adopt a Syncrude prototype policy in a bid to regain the nation's position of net energy self-sufficiency, the public will be carrying much of this cost—as well as the opportunity cost of what must be foregone or sacrificed in terms of other pressing social and economic needs—with little guarantee that the real owners of the resources will recoup anything approaching their true value. Further, a policy of providing incentives to the large international energy companies in the hope that they will increase supplies is sure to add to the country's long-run foreign ownership burden which is already intolerably high.

Will such a policy avert a confrontation with the United States over Canadian energy exports? Ottawa appears to be exceedingly worried about its energy relations with the U.S., and it has every reason to be. Our past decisions are now weighing heavily on our present choices, our freedom to manoeuvre is sharply constrained by the continental energy arrangements already in place. The U.S. appears to be adopting a "dollars for resources" position that, in effect, implies the threat to withhold capital, technology and expertise unless Canada agrees to continue supplying energy across the forty-ninth parallel—even if this means growing shortages for Canadians. The refusal of the U.S. Export-Import Bank to underwrite loans for Syncrude may well have been intended to flash this message to Ottawa, protestations to the contrary notwithstanding. In view of the rising demand in Canada for domestic energy and the virtual impossibility of rapidly developing reserves surplus to national requirements, it is hard to see how a confrontation with the U.S. over future oil and gas exports can be avoided without inflicting serious hardship on Canadian consumers. Certainly, Canada is going to be hard-pressed to attain its declared objective of national self-sufficiency in energy while increasing its dependence on a wealthy, powerful American industry that historically has been virtually inseparable from U.S. diplomatic and security policy. Canadians too must begin to understand that the security and strategic aspects of energy resources are likely to become increasingly critical in the years ahead. Issues that once could be left in the hands of businessmen and government economists now impinge directly on the broader question of Canada's sovereignty and security. If war is too important a business to be left to generals, energy is too important a business to be left to oil men.

Canada should now begin to move toward a much more activist energy policy that has the dual objective of (a) improving our domestic supply and demand equation, while (b) at the same time reducing our excessive dependence on the U.S. and its oil industry. Ottawa's current policies stress the creation of new supplies (next to nothing has been done to cut demand), but they also stimulate, indeed underwrite, the rapid growth of foreign ownership and control of the nation's untapped energy resources. Such an approach is shortsighted and inherently contradictory, since neither the U.S. government nor the international oil industry has any interest in seeing Canada attain a position of energy self-reliance and independence. At best, it will postpone hard choices and compound the security risk that goes with our present energy stance.

The problems are not simple, however, and there are few ready made panaceas at hand. There is little evidence, for example, that either Ottawa or the Canadian public is ready for drastic solutions such as the wholesale nationalization of the oil industry. This sort of measure could rapidly bring about a major confrontation with the United States and a full-blown economic and political crisis for which Canadians have not been prepared might well ensue. The likely result of such a policy, prematurely implemented, could be a political debacle of the first order and a sharp reversal for those who support the idea of Canadian independence. The day may soon come when such dramatic remedies seem both necessary and inevitable to large numbers of Canadians. But that day, unfortunately, is not here now. Then what is to be done?

As an alternative to across-the-board expropriation on one hand, and a continuation of present policies on the other, Canada could begin to move on several fronts to implement the two objectives of improving energy supply while enhancing national sovereignty. At home, the veto power of the multinational companies over future energy development must be ended, and the state must move aggressively into ownership of selected resources. Abroad, Canada should shift away from energy bilateralism to a policy emphasizing independence and multilateral approaches.

Although nationalization of the entire oil industry is plainly not a realistic political option now, Ottawa must nevertheless reconsider its position on public ownership of key resources and projects. Ottawa should view the issue of public ownership from the perspective of bargaining power, not of ideology. As has been demonstrated in the study of Syncrude, the political effect of ruling out the public ownership option is to undermine one's own bargaining position and to give added leverage to the multinational companies that are presently sitting on our resources and looking for the best available terms. Canada must be willing to consider the public option, if only to improve its negotiating stance in bargaining with transnational firms and to avoid situations in which it finds itself being played off against other governments. But it is obvious that such a policy can only succeed if a government is prepared, in the final analysis, to establish its reputation for consistency and toughness by moving swiftly and aggressively into public ownership of certain key resources or individual projects when the industry begins its familiar pressure tactics of withholding capital and threatening scarcity. Such tactics should be treated as a threat to Canada's overall security and countered with a demonstration of resolve and strength. Ottawa should choose its own battleground—it might decide to take over a single large project, or it could move to bring one of the major oil companies, preferably Imperial Oil, under public ownership—and it would have to be prepared to see the matter through. What is essential is that the government of Canada signal through some dramatic move that it has the resolve and the public support to break the veto power of foreign companies over Canadian resources.

Provided—and it is no small proviso—that the extraordinarily difficult jurisdictional disputes with Alberta could be sorted out, the tar sands might be a useful place for Ottawa to begin this policy of limited public ownership. Government involvement in the tar sands has a long and respectable tradition: without that involvement the truth is that Canada's oil sands would not be in production today.

But, in spite of the tradition, for all practical purposes, it is the large resource companies that enjoy exclusive concessionary privileges and control over this and most other Canadian resources. "Effective entry," Eric Kierans noted in his report on Manitoba's resource policies, "is confined to the very rich and the very large corporations who are unwilling, given alternative possibilities in other countries, to offer reasonable returns to the people whose resources they have enclosed." By default, the companies holding leases in the tar sands hold monopoly power, the power to overcharge and to restrict supplies, and they are using this power to ensure that development either proceeds according to their terms or does not proceed at all. Minority public participation in a venture such as Syncrude does not change this basic situation, but it certainly does enlarge the margin of public costs and risks. The joint venture, or "partnership" approach, implies a rough equality of power and mutuality of interests between the multinational oil companies and our governments, but the power is not shared equally and their interests are not compatible in the long run. "Participation" may therefore be the least acceptable approach to new energy resource development.

It would be no impossible undertaking to develop the tar sands under a jointly owned federal-provincial crown corporation, with ownership rights to the resource remaining in the hands of Albertans. The advantages of doing this today at the outset of the development process are obvious, as are the benefits that would flow from being able to plan for the ecological problems with the illogical leasing system ended. The people of Alberta and the rest of Canada would be in a position to capture all the economic rents for little more—and possibly less—expense or risk than they are at present incurring. A large Canadian industry could be developed to build and service the plants; Canadian engineers and scientists could lead the way in developing the resource; and the country would be guaranteed a steady, growing supply of high quality energy. The costs of such a policy would be high in the initial years admittedly, and there would still be some economic risk and environmental damage associated with tar sands development—public ownership is not a panacea—but this approach would reduce our children's burden of foreign ownership and dependence and it would also mark a decisive political precedent. And this, of course, is why the oil industry and its powerful political allies would fiercely resist the attempt to bring a resource such as the tar sands under total public ownership and control.

On the international level, Canada should begin to balance its bilateral energy contacts and relations with the United States by adopting a more independent stance on international energy issues and by utilizing, where available, multilateral approaches in energy diplomacy. Official bilateralism must yield to a more activist defence of Canadian interests. This would be perfectly consistent with Ottawa's own "third option" in foreign policy, namely the gradual reduction of our vulnerability and dependence on the U.S. through the diversification of our political, economic, cultural and military relationships. To cite some areas where bilateralism should give way to a mix of independence plus multilateralism, Canada should publicly dissociate its international energy policies from those of the U.S. State Department and it should express its strong opposition to moves that are plainly intended to enhance American interests at the expense of resource producers in the Third World. Ottawa should not align itself with Washington in any diplomatic efforts to undermine and destroy OPEC or similar commodity associations. Such efforts are likely to be counter-productive, will earn Canada the enmity of developing countries, and also begin from the false premise that our interests necessarily coincide with those of the United States and its corporations. In truth, Canadian interests may often lie with the producers instead of the consumers. Canada should, however, try to mediate between producers and consumers, acknowledging the aspirations of many developing nations for a new international economic order; but such a policy would require a clear-cut break with the heavy-handed tactics that have often characterized U.S. diplomacy on resource matters. In the long run Canada's interests may lie in attempting to take a positive part in fashioning a more equitable, and also more stable, world economic order; certainly they do not lie in a policy designed to help recreate an American hegemony. An old order is passing; Canadians should acknowledge that fact and try to play a progressive role in assisting the birth of its successor.

A multilateral approach, as opposed to our traditional bilateralism, could well be in Canada's interest in certain other areas. Problems of Arctic sovereignty, northern resource development, offshore resource exploitation are obvious candidates for a multilateral approach in external policy. Other writers have advocated a closer association with countries such as Norway, Denmark and the Soviet Union in defending Canadian sovereignty in the Arctic from American encroachments. Canada has much to learn from—and perhaps something to teach—such countries, with whom she shares many common problems and interests in the north. Canada should also have better links with other oil producers—including Norway, the OPEC bloc, Mexico—that have bargained more effectively with the international industry. An exchange of information concerning the strengths and weaknesses of the various major companies, of world reserves, supply and demand estimates, future price trends, replacement costs and so on, would give Canada needed leverage in its negotiations with the oil industry and reduce the possibility of the companies returning to their old game of playing producers off against each other. Canada would not have to join OPEC to gain the advantages of such a policy but she would certainly have to make some kind of break with American energy strategy in order to win the confidence of other producers.

Much more could be added along these lines. But it must be understood that there are real options in energy open to Canadians between the extremes of Syncrude and wholesale nationalization. An activist mix of domestic and international energy policies, emphasizing public ownership of selected resources and projects at home and independence and multilateralism abroad could, if aggressively and consistently implemented, take Canada a long way toward the resolution of her growing energy predicament. Beyond this, it would also commence the necessary process of repatriating key areas of the Canadian economy and changing the conditions that gave rise to the politics of Syncrude.

Appendix   The Blair Report

One impediment we face in trying to discuss the economics of oil sands development lies in the lack of available data on what it costs to produce a barrel of synthetic fuel and what that barrel of fuel is worth. Sun Oil and Syncrude obviously have devoted much attention to these questions, but their answers have not been made available to the public.

Although its applicability to today's developments is obviously limited, the Blair Report of 1950[1] looked into these questions for the Alberta government in an attempt to show that commercial development was feasible at that time. The answers contained in the report are quite possibly of more than purely historical interest.

Blair visualized a fully-integrated surface mining, extraction and processing plant producing 20,000 barrels per day on one square mile of deposit in the Mildred-Ruth lakes area, near the present sites of GCOS and Syncrude, for a fifteen year period.

The Blair Report broke down the costs of tar sands production as follows:

Mining	*$/bbl. of synthetic fuel* *("desulphurized distillate")*
The cost of overburden removal, excavation, conveying, stock-piling of wastes.	$0.55
Processing	
1) Hot water extraction & coking	0.72
2) Hydrogenation	0.81
Total	$1.53

Pipeline to Edmonton	0.28
Pipeline Edmonton to market at the Great Lakes	0.55
Allowances for storage and other expenses	0.19
Total	1.02
Total Costs	$3.10
Estimated Market Price	3.50
For a Profit of	$0.40

And a Return of 5.5 percent on capital invested.

The value of what Blair called the "desulphurized distillate" was estimated in two ways. First, by determining what could be made from the distillate and what these products were worth. This indicated that a barrel of distillate would yield products worth $4.03 at a Chicago refinery. Second, market price statistics were used to evaluate the value of the distillate streams (basically, naptha and furnace oil), and on this basis Blair came up with a price one-third higher than the crude oil price. Alberta Redwater crude was at that time valued at $3.00 per barrel (delivered to market) and Blair conservatively set the value of the distillate at one-sixth of that or $3.50.

It is also intriguing to note that discussions at the 1951 Oil Sands Conference indicated that Blair's estimated costs were too high and his price too low. "It is probable that costs for really large plants would be considerably less," Blair summed up.

It was on the basis of these calculations that Blair concluded in 1950 that commercial exploitation of the Alberta tar sands was feasible—immediately.

[1] Blair, S. M., in association with Bechtel Corporation and Universal Oil Products Company, "The Development of the Alberta Bituminous Sands," Board of Trustees, Oil Sands Project. Government of Alberta, Edmonton. December 1950.

Notes

1. In a 1973 study of the Athabasca deposit the Energy Resources Conservation Board of Alberta estimated "potential surface mineable reserves" at 74 billion barrels of crude bitumen and "potential *in situ* reserves" at 686 billion barrels. The board defines "proved recoverable reserves" as those deposits containing five percent or more by weight of bitumen, lying under less than 150 feet of overburden, and with an overburden to pay ratio of less than one (i.e., depth of tar sands exceeds that of the overburden). These assumptions led the ERCB to conclude that "the total recoverable reserve now considered proved by mining methods is 38 billion barrels of crude bitumen." Under current separation and conversion processes one barrel of crude bitumen yields about 0.7 barrels of synthetic oil, thus the ERCB concluded that 26.5 billion barrels of synthetic oil could be designated "proved recoverable."

2. At the end of 1974 it was revealed that the U.S. Export-Import Bank had refused a loan of about $75 million to the Syncrude group. Interestingly, this announcement came just days after the release of a National Energy Board Report which called for the phasing out of oil exports to the U.S. by 1982 (see chapter ten). The U.S. government wants to see the tar sands developed, but for a continental market.

3. GCOS has reportedly been given special permission on a number of occasions to dump effluent wastes into the Athabasca River and to exceed provincial standards on atmospheric emissions of sulphurous wastes.

4. Peter Lougheed later recalled that in 1972 he had been told by Canadian Petroleum Association officials that Canadian oil prices should rise by about ten cents a barrel per year until 1980. In 1972-73 they rose by ninety-five cents before Ottawa froze prices. This encouraged him, he noted, to raise oil royalties in 1973 and 1974.

5. In September 1974 the first commercial shale project, the Colony Project, was shelved indefinitely by a group made up of Atlantic Richfield and

several other companies. Since that time there has been little or no activity in the Colorado shales.

6.   Non-deductibility of royalties at a fifty percent tax rate would cost $928 million. But the budget intended to cut the tax rate by forty percent, reducing that cost by $186 million and reducing by $112 million the tax burden on the increases in revenues net of royalties: W. Gainer and T. Powrie, "Public Revenue from Canadian Crude Petroleum Production." *Canadian Public Policy*, Winter 1975.

7.   "Producibility is the estimated average annual ability to produce, unrestricted by demand but restricted by reservoir performance, well density and well capacity, oil sands mining capacity, field processing and pipeline capacity"; National Energy Board, *In the Matter of the Exportation of Oil*, October 1974.

8.   For example, Amoco Canada (Standard of Indiana) has suggested that the technology is available to double the amount of Alberta's remaining proven oil reserves that can be produced. Exotic or third generation recovery methods could produce a "possible" 6.3 billion barrels of oil generally not regarded as recoverable, given special incentives, Amoco engineers have argued. See the Edmonton *Journal*, April 25, 1975.

9.   Not only the oil companies want "partnership." Leaked documents relating to Alberta's proposed world scale petrochemical complex show that chemical giant Dupont is insisting that the Alberta Energy Company have a direct equity involvement in the project.

10.   In the last week of January both Alberta and Ottawa agreed that they would require a month to six weeks to digest the findings of the consultants' reports. As it turned out, they did so in a matter of hours: none of the reports were completed and in the politicians' hands before the expiration of the deadline. The key report, that of Loram International on the huge cost overruns, has not been made public in full. A summary of the report notes that "it is not possible to completely reconcile the two estimates" of July 1973 and December 1974, yet "there are justifiable reasons for the increases in cost." These were (1) severe, unanticipated inflation; (2) additional construction costs; (3) estimate revisions as the engineering definition of the project advanced. These general assertions tell us next to nothing, and without the full Loram report it is impossible to take its conclusions at face value.

11.   Assuming in both cases a federal corporate tax rate of twenty-five percent and a provincial rate of eleven percent. After deducting these tax concessions of $670 million the maximum net outlay of the companies is $528 million, or about twenty-five percent of capital costs for seventy percent ownership. The public sector's commitments of equity, tax concessions and loans total $1.472 billion, or seventy-five percent of the cost for thirty percent of the ownership and control.

Select Bibliography

Adelman, M. A. "Is the Oil Crisis Real?" *Foreign Policy,* Winter, 1972-3.

Alberta Conservation and Utilization Committee. *Alberta Oil Sands and Meteorological Research.* Edmonton: Alberta Environment, March 1974.

Alberta Oil Sands Reclamation Research. Edmonton: Alberta Environment, March 1974.

Fort McMurray Athabasca Tar Sands Development Strategy. Mimeographed. August 1972.

Alberta Oil Sands Hydrological Research. Edmonton: Alberta Environment. March 1974.

Alberta Department of Federal and Inter-Governmental Affairs. *The Alberta Oil Sands Story.* January 1974.

Alberta Department of Industry and Commerce. *Fort McMurray Community Profile.* 1973.

"Management of Growth." 29 May 1974.

Alberta Department of Mines and Minerals. "Federal Government Abasand Oils Ltd." 28 May 1974.

Alberta Oil and Gas Conservation Board. *A Description and Reserve Estimate of the Oil Sands of Alberta.* 1963.

Alberta Oil Sands Corridor Study Group. "Alberta Oil Sands Corridor Study." Mimeographed. June 1974.

Alberta Provincial Auditor. *Great Canadian Oil Sands Ltd.: Report on Royalty Audit and Special Study.* 6 December 1974.

Alberta Select Committee of the Legislative Assembly on Foreign Investment. *Final Report on Foreign Investment.* December 1974.

Alberta. "Final Report of the Interdepartmental Working Group on the Oil and Gas Contingency Plan (amended version)." 22 October 1974.

Alberta. *Oil Sands Development Policy Considerations.* June 1974.

"Alberta Oil Sands and Mining Report" *Oilweek* (17 March 1975).

Atlantic Richfield Canada Limited; Canada-Cities Service Limited; Gulf Oil

Canada Limited; and Imperial Oil Limited. "An Application to the Energy Resources Conservation Board to Amend Approval No. 1223 of the Oil and Gas Conservation Board." Mimeographed. August 7, 1971.

Barnet, Richard J. and Müller, Ronald E. *Global Reach: The Power of the Multinational Corporations*. New York: Simon and Shuster, 1974.

Berry, Glynn R. "The Oil Lobby and the Energy Crisis." *Canadian Public Administration* (Winter 1974).

British Petroleum. "Energy in the Future." *BP Shield International,* December 1973.

Camps, F. W. "Tar Sands," *Encyclopedia of Chemical Technology* (2nd Ed.).

Canada. Minister of Energy, Mines and Resources. *An Energy Policy for Canada-Phase 1*. Volume 1, Ottawa: Information Canada, 1973.

Canadian Petroleum Association. "Submission to the National Energy Board." Mimeographed. December 1973.

Canadian Petroleum Association and Independent Petroleum Association of Canada. *The Petroleum Industry and Canada*. September 1974.

Carrigy, M. A. and Kramers, J. W. "Geology of the Alberta Oil Sands." Paper presented at the Western Region Tar Sands Conference, Edmonton, 17-19 April, 1974.

Carrigy, M. A. Comp. *Athabasca Oil Sands Bibliography (1789-1964)*. Preliminary Report 65-3. Edmonton: Research Council of Alberta, 1965.

The K. A. Clark Volume; a Collection of Papers on the Athabasca Oil Sands Presented to K. A. Clark. Information Sines 45. Edmonton: Research Council of Alberta, 1962.

Carrigy, M. A. and Kramers, J. W., eds. *Guide to the Athabasca Oil Sands Area*. Edmonton: Alberta Research, 1973; reprinted ed., 1974.

Central Mortgage and Housing Corporation. "Fort McMurray: New Town Status." Mimeographed. April 1965.

Clark, K. A. "Athabasca Bituminous Sands." *Fuel* 30 (1951): 49-53.

"Commercial Development Feasible from Alberta's Bituminous Sands." *Canadian Oil and Gas Industries* (October 1951). Reprint.

Clegg, M. W. "New Sources of Oil—Oil Sands, Shales and Synthetics." In *Energy: From Surplus to Scarcity*. Edited by K. Inglis, New York: John Wiley and Sons, 1974.

Comfort, D. J. *Meeting Place of Many Waters: A History of Fort McMurray*. Fort McMurray: Comfort Enterprises, 1973.

Ells, S. C. *Recollections of the Development of the Athabascan Oil Sands*. Information Circular 1C-139. Department of Mines and Technical Surveys, Government of Canada, Ottawa.

Engler, Robert. *The Politics of Oil: Private Power and Democratic Directions*. Chicago: University of Chicago Press, 1961.

Environment Canada. "Memorandum and Correspondence Relating to the Syncrude Environmental Impact Assessment." September 1974.

Erickson, Edward W. and Waverman, Leonard, eds. *The Energy Question: An International Failure of Policy*. 2 Vols. Toronto: University of Toronto Press, 1974.

Fitzsimmons, R. C. *The Truth About Alberta Tar Sands: Why Were They Kept Out of Production.* Edmonton: 1955.

Foster Economic Consultants Limited. "Calculation of Albertans' Share of Profits from Syncrude Projects." September 1973.

"Principal Risk Areas of the Syncrude Project." September 1973.

Foster Research Limited. "Economic Evaluation of the Syncrude Project." January 1975.

Freeman, J. M. *Biggest Sellout in History: Foreign Ownership of Alberta's Oil and Gas Industry and the Oil Sands.* Edmonton: Alberta N.D.P., 1966.

Gainer, W. D. and Powrie, T. L. "Public Revenue from Canadian Crude Petroleum Production." *Canadian Public Policy* (Winter 1975).

Government of Canada. "Correspondence between the Minister of Finance and Imperial Oil Ltd. and Atlantic Richfield Canada Ltd., December 5, 1973 and January 24, 1975."

Govier, G. W. "Alberta's Oil Sands in the Energy Supply Picture." A talk to the C.S.P.C. Symposium on "Oil Sands—Fuel of the Future." Calgary, 1973.

Gray, Earle. *The Great Canadian Oil Patch.* Toronto: Maclean-Hunter, 1970. Chapter 14.

Hanson, Eric J. *Dynamic Decade.* Toronto: McClelland and Stewart, 1958.

Hanson, James W. "Energy: A Global Perspective." Paper presented to the Conference Board in Canada, Vancouver, 30 May 1974.

Imperial Oil Ltd. "Submission to the National Energy Board in the Matter of the Exportation of Oil." Mimeographed. December 1973.

Intercontinental Engineering of Alberta Ltd. *An Environmental Study of the Athabasca Tar Sands.* Edmonton: Alberta Environment, March 1973.

Laxer, James. *The Energy Poker Game: The Politics of the Continental Resources Deal.* Toronto: New Press, 1970.

Canada's Energy Crisis. Toronto: James Lewis and Samuel, 1974.

Laycock, A. H. "Water Problems in Alberta Oil Sands Development." Mimeographed. 30 January 1975.

Loram International Limited. "Executive Summary of Conclusions and Findings." 4 February 1975.

Lynnett, Steve. "Cold Lake." *Imperial Oil Review.* 58 (1974): 2-7.

"Digging for Oil." *Imperial Oil Review.* 57 (1973): 16-23.

Mathias, Philip. *Forced Growth.* Toronto: James Lewis and Samuel, 1971.

Medvin, Norman. *The Energy Cartel: Who Runs the American Oil Industry.* New York: Vintage Books, 1974.

Moore, S. Donald. "Nuclear Energy as a Subsurface Heavy Oil Recovery Technique." Paper presented at the Western Region Tar Sands Conference, Edmonton, 17-19 April, 1974.

National Energy Board. *Report to the Honourable Minister of Energy, Mines and Resources in the Matter of the Exportation of Oil.* October 1974.

O'Conner, Harvey. *World Crisis in Oil.* New York: Monthly Review Press, 1962.

The Empire of Oil. New York: Monthly Review Press, 1962.

Odell, Peter, R. *Oil and World Power: Background to the Oil Crisis.* Hammondsworth: Penguin Books, 1974.

Pasternack, D. S. "Alberta Oil Sands." *The Petroleum Engineer,* February 1953. Reprint.

Petrofina Canada Ltd. "A Submission to the National Energy Board in the Matter of the Exportation of Oil." Mimeographed. December 1973.

Reid, Crowther and Partners Ltd. *Report on the Impact of a Proposed Synthetic Crude Oil Project on Fort McMurray.* Edmonton: Syncrude, February 1973.

Ridgeway, James. *The Last Play: The Struggle to Monopolize the World's Energy Resources.* New York: Mentor, 1973.

Robbins, Sidney M. and Stobaugh, Robert B. *Money in the Multinational Enterprise: A Study in Financial Policy.* New York: Basic Books, 1973.

Rowland, Wade. *Fuelling Canada's Future.* Toronto: Macmillan, 1974.

Ryan, J. T. "Supply and Demand for Crude Oil in Canada." Paper presented at the Western Region Tar Sands Conference, Edmonton, April 1974.

Scott, Anthony. *Natural Resources: The Economics of Conservation.* Toronto: McClelland and Stewart Ltd. 1973.

Shell Canada Ltd. "Submission to the National Energy Board in the Matter of Exportation of Oil." Mimeographed. December 1973.
    "Application of Shell Canada Ltd. and Shell Explorer Ltd. to the Energy Resources Conservation Board." Mimeographed. June 1973.

Spragins, F. K. "What Role will the Athabasca Tar Sands Play in Solving Canada's Energy Crisis." Address at the Western Region Tar Sands Conference, Edmonton, 17 April 1974.

Sun Oil Company Ltd. and Great Canadian Oil Sands Ltd. *Submission to the National Energy Board in the Matter of the Exportation of Oil.* December 1973.

Syncrude Canada Ltd. *Basic Information on the Syncrude Project.* Edmonton: Syncrude, 1973.
    Syncrude Lease No. 17: An Archaelogical Survey. Environmental Research Monograph, 1973-74.
    1973: Year of Decision.

Syncrude Project Agreements of September 14, 1973 and, as amended, December 13, 1973 between the Province of Alberta and Imperial Oil Ltd., Gulf Oil Canada Ltd., Canada-Cities Service Ltd. and Atlantic Richfield Canada Ltd.

Science Council of Canada. *Canada's Energy Opportunities.* Report No. 23. Ottawa: Information Canada, 1975.

Shaffer, E. H. *The Oil Import Program of the United States.* New York: Praeger, 1968.

Tanzer, Michael. *The Political Economy of International Oil and the Underdeveloped Countries.* Boston: Beacon Press, 1969.
    The Energy Crisis: World Struggle for Wealth and Power. New York: Monthly Review Press, 1974.

"The Syncrude Project: To Satisfy the Energy Thirst." *Esso Reporter* (June 1973)

Thomas, Gene. "Athabasca Tar Sands—Future Energy Source." *The Texaco Star*. 59 (Summer 1974): 18-23.

Thiessen, H. W. "The Impact of the Oil Sands Development." Paper presented at the Western Region Tar Sands Conference, Edmonton, 17-19 April 1974.

Tottrup Engineering Ltd.; Kates, Peat, Marwich and Co. Ltd.; and F. F. Slavey and Co. Ltd. "Athabasca Tar Sands Gathering System Study." Mimeographed.

Tugendhat, C. *Oil; the Biggest Business*. London: Eyre and Spottiswoode, 1968.

U. S. Federal Trade Commission, *The International Petroleum Cartel*. (1952).

Walwyn, Stodgell and Co. Ltd. *A Look at World Energy: The Athabasca Tar Sands*. June 1973.

Watson, J. F. *Oil from Non-Conventional Sources*. London: British Petroleum, May 1972.

W. J. Levy Consultants Corporation (New York). "Economic Considerations Relevant to Canadian Oil Policy." Mimeographed. December 1973.

Emerging North American Oil Balances: Considerations Relevant to a Tar Sands Development Policy: Appendices. Mimeographed. February 1973.

Wright, R. W. and Mansell, R. L. "The Impact of Large Scale Investment on the Alberta Economy and the Role of Migration in the Adjustment Process." Mimeographed.

Yurko, W. J. "Development of the Alberta Oil Sands." Address to the Western Region Tar Sands Conference, Edmonton, 17 April 1974.

"Environmental Problems in the Development of the Fort McMurray Tar Sands." An address to the Instrument Society of America, Calgary, 23 January 1973.